Contents

For Sam

Tractors since 1889

45101

Tractors since 1889

Michael Williams

Special photography by David Williams

FARMING PRESS

First published 1991

A catalogue record for this book is available
from the British Library

ISBN 0 85236 223 4

(*Frontispiece*)
A trainload of Oliver tractors made in the David Brown factory in England on its way to America in 1960.

Dust jacket design by Andrew Thistlethwaite

Published by Farming Press Books
4 Friars Courtyard, 30–32 Princes Street
Ipswich IP1 1RJ, United Kingdom

Distributed in North America
by Diamond Farm Enterprises,
Box 537, Alexandria Bay, NY 13607, USA

Phototypeset by Galleon Photosetting, Ipswich
Printed and bound in Great Britain by
Butler and Tanner Ltd, Frome, Somerset

Acknowledgements

I WOULD like to thank everyone who helped in the preparation of this book, and particularly those who provided photographs and information or allowed their tractors to be photographed. Thanks also to Judith for helping with translation from French and German.

The following photographs were taken by David Williams: colour plates 2, 5, 6, 7, 8, 10, 11, 15, 16, 18, 19, 20, 21A, 23, 24, 30 to 47 and the photographs used on the back cover, plus the black and white photographs on pages 29, 33, 39, 46, 66, 71, 73 upper, 76, 82 and 83. The copyright to the photographs on pages 36, 44 lower and 57 is held by the Museum of English Rural Life, and British Motor Industry Heritage Trust supplied the picture on page 97. Photographs on pages 15, 63 and 94 are from the Collections of Greenfield Village and the Henry Ford Museum, and the photograph on pages 40–41 is from the Idaho Historical Society.

David Gaines supplied the frontispiece picture, Deere and Co supplied the front cover picture and photographs on pages 10, 20–21, 68–69 and 79 upper, and colour plates 21B, 22, 25, 26 and 29 are from Walterscheid. Aebi, Allis-Chalmers, Case IH, Citröen, Daimler-Benz, Deutz, Ford New Holland, Hungarian Agicultural Museum, JCB, Klöckner-Humboldt-Deutz, Kubota, Landini, Ransomes, Renault, SAME, Steyr-Daimler-Puch and Vickers also provided photographs used in the book, and other photographs are from the author's collection.

Michael Williams

1

One Hundred Years of Tractors

TRACTOR history started more than one hundred years ago in the United States. The first tractor was probably the Charter made at Sterling, Illinois in 1889 by the Charter Gas Engine Co run by John Charter.

He mounted a heavy, single cylinder Otto petrol engine on a chassis he made up from the wheels and transmission of a steam traction engine. The tractor was sent to a farm near Madison, South Dakota, where it was used to power a threshing machine. The results must have been satisfactory, as the Charter company received orders for a further five or six tractors based on the same design.

Charter tractors were soon facing competition. In 1892 the J I Case Threshing Machine Co of Racine, Wisconsin built their first prototype tractor, and in the same year tractors were also built by John Froelich in Iowa and by the Dissinger brothers of Wrightsville, Pennsylvania.

Case prototype tractor completed in 1892.

None of these tractors achieved much commercial success initially. They were all designed mainly for stationary work, and they were based on steam engine type running gear with a stationary engine to provide the power.

The Case tractor used a twin cylinder Patterson engine, but problems with the ignition and fuel supply persuaded Case to abandon the tractor project and concentrate on their highly successful steam engines, which were much more reliable. Case returned to the tractor market almost 20 years later to become one of the leading manufacturers.

The first Froelich tractor was based on a Robinson chassis and a Van Duzen vertical engine with a big single cylinder of 14 in. bore and stroke. The tractor was sent to South Dakota to work with a threshing machine, and the results encouraged John Froelich to establish the Waterloo Gasoline Traction Engine Co based at Waterloo, Iowa. Much later, in 1918, the Waterloo company was taken over by Deere and Co, and the 1892 Froelich tractor is described as the forerunner of the John Deere tractor range.

A replica of the 1892 Froelich tractor, an early ancestor of the John Deere range.

This 15hp tractor was made in about 1896 by the American Otto Company.

The Dissingers of Wrightsville, Pennsylvania equipped their Capital tractor with an American made Otto engine. The tractor made little commercial impact at first, but Capital tractors became well known in the early 1900s.

More companies decided to move into the tractor market during the 1890s including the Otto Gas Engine Works, the American subsidiary of the German Deutz company. The Huber company of Marion, Ohio began making tractors in 1898, using their own steam engine chassis and Van Duzen engines. Huber was probably the most successful of the pre-1900 tractor companies and remained prominent in the tractor industry until the late 1930s.

The tractor market in America and Canada expanded quickly in the early 1900s as manufacturers began to design tractors for

(*Facing page*)
(*Above*)
The first Hart-Parr tractor was built in 1901 at Charles City, Iowa.

(*Below*)
The 40–70 Imperial tractor weighed 10½ tons.

ploughing. This meant providing a stronger transmission, and it also required bigger engines and additional weight to help overcome traction difficulties. The result was a new generation of tractors which were impressively big and heavy, but also expensive.

There were literally dozens of different makes and models, and the best known examples included some of the early Rumely, International, Avery, Case and Hart-Parr models, Flour City, Twin City, Gaar Scott, Nichols and Shepard, Big Four and Pioneer. Imperial was one of the less successful makes, with a 10½ ton 40–70 model powered by a horizontally opposed, four cylinder engine, which was available until about 1919. The Imperial was known in Britain as the Goliath tractor.

The first commercially successful British tractor, the Hornsby-Akroyd Patent Safety Oil Traction Engine from Grantham, Lincolnshire, was also a heavyweight. It made its first appearance in 1896 and was probably the first tractor anywhere with an engine designed to burn paraffin or kerosene instead of petrol. The single cylinder Hornsby-Akroyd engine worked on the hot bulb principle,

Hornsby-Akroyd tractor production started in 1896.

Harvesting with an Ivel tractor in about 1905.

needing a blowlamp to create a hot spot in the cylinder head to start the engine from cold. Once the engine was running, the blowlamp could be switched off as heat from the combustion chamber maintained the working temperature.

The Hornsby-Akroyd was also the first tractor to be sold in Britain—the customer owned an estate near Weybridge, Surrey—and it was probably also the first tractor to win an export order when several were shipped to Australia during the period 1897 to 1899.

Other British companies also built big tractors from about 1907, particularly Marshall which developed substantial export trade to countries with big farms, such as Australia and Canada.

Most of the British manufacturers in the period up to 1920 concentrated on making lighter, cheaper tractors. Companies such as Ivel, Scott, Ransomes and Saunderson led the world in the development of tractors which made a complete break from traction engine ideas and were the real forerunners of modern, multi-purpose tractor power.

The Ivel tractor designed by Dan Albone of Biggleswade, Bedfordshire was easily the most successful of the early tractors in Britain. Ivels were exported to more than 20 countries, and were also made in the United States under a licence agreement.

Early versions of the Ivel made from about 1903 weighed 1500 lb, which attracted favourable comment from farmers concerned about soil compaction. The power unit was a twin cylinder

Allis-Chalmers 10–18 tractor displayed in the Henry Ford Museum in Dearborn, Michigan.

petrol engine running at a relatively fast 850 rpm, and the tank for the cooling water was placed over the rear axle where the weight would be most useful for traction.

British farmers were not ready for the tractor until the First World War had transformed the economic situation and had removed much of the manpower from their farms. Dan Albone's efforts to build up a demand for his tractors were ingenious and energetic, and they included regular demonstrations on a local farm where he kept a tractor and a collection of machinery. He also looked for alternative markets, and he demonstrated an Ivel tractor as a fire engine with a high pressure water pump; he also made a bullet-proof version which he demonstrated as a military ambulance to an apparently unenthusiastic War Office.

Dan Albone died in 1906, and without his leadership and ideas the company gradually faded into obscurity.

Meanwhile American manufacturers were beginning to design

tractors suitable for smaller farms, and these began to sell in large numbers as the war in Europe created a new export market. Dozens of companies began making smaller tractors, and just a few of these became really successful.

The success stories included the Little Bull tractor, which arrived on the market in 1914 and moved towards the top of the best seller lists, in spite of its curious design with three different sized wheels, and in spite of providing only 5 hp at the drawbar. A more powerful version, known inevitably as the Big Bull, attracted the attention of the Massey-Harris company which was keen to move into the tractor market, and they signed a marketing deal to sell the Big Bull in Canada (see my book *Massey-Ferguson Tractors*).

Allis-Chalmers decided to become a tractor company in about 1912, and with considerable foresight they ignored the fashionable heavyweight end of the market and developed a medium

Allis-Chalmers General Purpose motor plough.

Martin's of Stamford motor plough with tracks.

sized tractor, the 10–18, which weighed two tons and was available from 1915. The horizontally opposed twin engine started on petrol and ran on paraffin.

Although demand for the 10–18 proved to be disappointing, Allis-Chalmers remained committed to the small tractor market and announced the General Purpose model in 1918. This was a motor plough, a special type of tractor which was popular for about ten years from 1915.

Motor ploughs provided many farmers with their first opportunity to switch from animal power. When Allis-Chalmers moved into the market there were still an estimated 25 million horses and mules on American farms, suggesting an enormous potential market for a low cost tractor.

The Allis-Chalmers version was powered by a Le Roi four cylinder engine with a 6–12 power rating (6 hp at the drawbar and 12 hp at the flywheel or belt pulley). Production reached 700 when the last of the motor ploughs was built in 1922.

Other manufacturers were more successful, and the Moline Universal was probably the best selling of the American motor ploughs. It was also the most advanced, and later versions were equipped with an electric starter, probably the first time this feature was available on a tractor.

The motor plough fashion also spread to Europe, and particularly to Britain where imports from America competed against home produced machines from a long list of companies including Martin's of Stamford, Fowler and the Crawley made by two brothers who farmed near Saffron Walden, Essex.

In Austria and Czechoslovakia the motor plough became big and powerful, a departure from the original cheap-and-cheerful concept developed in America. The Excelsior motor plough was made by the Puch organisation, one of the forerunners of the Steyr Daimler Puch group which makes Steyr tractors in Austria. The engine on the 1919 Excelsior in the photograph developed a maximum output of 40 hp from seven litres capacity, and the three ratio gearbox provided a maximum forward speed of 2.8 mph.

Motor ploughs vanished from the market during the early 1920s. Their fate was sealed in 1917 when the first of Henry Ford's Model F Fordson tractors came off the production line at Dearborn, Michigan following an urgent request from the British Government.

The Fordson had some limitations. It tipped over backwards too easily and traction was poor in difficult conditions, but it was sturdy and reasonably reliable, and it was the first tractor designed for mass production. No other tractor since 1889 has had more impact on farm mechanisation, and none has achieved such a spectacular success, with production reaching almost 750,000 before the first major design changes were made in 1928.

(Facing page)
(Above)
This Puch Excelsior motor plough was powered by a 7 litre petrol engine in 1919.

(Below)
Model F Fordson exported to Switzerland in about 1925.

The Crawley motor plough was made in Essex and was known as an Agrimotor.

Helvetia.

JOHN DEERE

The most compelling reason why so many farmers chose a Fordson was the price. The Allis-Chalmers motor plough sold for $850 in 1918 when the Fordson price was $750. Fordson prices were reduced repeatedly to reach only $395 in 1922, and in the same year Allis-Chalmers had to cut their prices to $295 to clear their unsold stock of motor ploughs.

Fordson's success, detailed in *Ford and Fordson Tractors,* put many rivals out of business, including the Model M Samson which General Motors had brought on to the market in 1918 to compete with Henry Ford's tractor.

The Fordson, plus the economic problems in the 1920s and early 1930s, started a rationalisation process in the tractor industry which is still continuing in the 1990s. Among the few really successful mid-1920s American competitors for the Fordson were the Farmall rowcrop tractors from International Harvester and the mechanically simple, two cylinder John Deere tractors.

Britain's tractor industry expanded rapidly after the end of the war in 1918, with new arrivals including Blackstone, Austin, Alldays and Onions, and the Fordson look-alike from Rushton. These, plus the few survivors from the pre-war industry including Saunderson, all pulled out of the tractor market as sales slumped. By 1930 the only survivor still in production was the Austin tractor, which was built in France; British production had ended in about 1923.

There was a similar situation throughout Europe, where most

(*Preceding pages*) Model D John Deere, one of the survivors of Fordson competition in the 1920s.

The Saunderson Universal tractor was Britain's best selling tractor until 1918.

The Nuffield Universal tractor was one of the new arrivals in the British tractor industry in the late 1940s.

of the 1918 to 1920 newcomers had disappeared by 1930. Exceptions included Deutz and Lanz in Germany and the Renault tractor range in France.

Another world war helped to revive the fortunes of the tractor industry. Production in Britain was boosted when the leading North American companies, including Allis-Chalmers, International Harvester, Massey-Harris and Minneapolis-Moline, began building tractors in British factories for the expanding European market. John Deere wanted to come to Britain, but chose Germany instead after disagreeing with the British Government over the location for a new factory.

There was a further boost when the Nuffield organisation moved into the market for the first time, and Harry Ferguson set up his TE tractor production line in Coventry to build Ferguson System tractors for the world.

After 30 years or so of varying degrees of prosperity, the tractor industry faced another period of rationalisation. Many of the most famous makes disappeared from the market or lost their individual identity through takeovers and mergers, and the latest of these in 1991 was the effective takeover of the Ford tractor and farm machinery operation by the Fiat group.

JCB

(Facing page)
JCB high speed tractor built in 1991.

In spite of the rationalisation, there has also been considerable technical development, including transmission improvements, quieter, more comfortable cabs, and electronic information and control systems.

We also appear to be on the brink of major design changes with the launch of the JCB high speed tractor in 1990 with a 45 mph top speed. This is a logical development from tractors with a mid-mounted cab pioneered by the Mercedes-Benz MB-trac and the Deutz Intrac from Germany.

2

Wheels and Tracks

THE first tractors were used mainly as stationary power units for driving threshing machines and other farm equipment, and pulling power was not one of the most important priorities. However, the situation changed quickly as farmers and contractors wanted to use tractor power for ploughing, pulling trailers and other jobs that demand traction.

Traction is the ability to convert engine power into pulling power. If the tractor wheels or tracks can grip the soil effectively, more of the engine power is available for work and the tractor can pull a bigger load or plough more acres in a day.

Ideas for improving traction began to appear in the early years of tractor development. Flat metal plates attached around the circumference of the driving wheels were often used on steam traction engines, and these became the standard traction aid for tractors.

Flat pieces of metal on the wheels meant that the tractor could be used on the road as well as on the land. The grip they provided was sufficient to deal with the relatively low power available from the early tractors, which were usually heavy enough to produce reasonably efficient traction in most soil conditions.

The development of more powerful tractors emphasised the need for better traction, and the increasing popularity of smaller, lighter tractors also helped to highlight wheelslip problems.

The link between traction and weight shows up clearly in wheelslip figures recorded during the University of Nebraska tractor tests. The number of tractors tested during 1920 was 65, and the heaviest of these was the Rumely Oil Pull 30–60 weighing 26,000 lb and achieving 5.22 per cent wheelslip in the ten hour drawbar test. Other heavyweights included the 40–65 Twin City, which weighed 25,550 lb and recorded 9.8 per cent slip, and the 24,450 lb Aultman-Taylor 30–60 which achieved 2.0 per cent wheelslip, one of the best results in the 1920 tests.

Wheelslip figures were generally higher for the lighter tractors taking part in the tests. The slip figure recorded for the 10–18 Case weighing 3760 lb was 17.6 per cent, a 3500 lb Wallis 15–25 produced 16.35 per cent slip and the 2710 lb Fordson, one of the lightest tractors tested in 1920, produced almost 24 per cent wheelslip—which was the worst result of the year.

One way to deal with the problem was to provide cleats, spuds or lugs which could penetrate deeper into the soil to provide a more positive grip. These improved the pulling power of tractors on soft ground, but they could not be taken on a surfaced road. To make the tractors roadworthy, special rims could be attached to provide a smooth surface to minimise damage.

'Mud wheel' for an Australian Ronaldson Tippett tractor in the early 1930s.

Farmers on heavy clay soils faced—and still do face—the biggest traction problems. Wheelslip is difficult to avoid in wet conditions, and soil sticking to the wheels reduces the effectiveness of the cleats. Several manufacturers developed special wheels to deal with the problem, and one example was the self-cleaning system used by the British company, Bumsted and Chandler on their Ideal tractor.

The spuds on the driving wheels of the Ideal were linked to a cam. As the wheels turned, each spud was pushed out through

Ferguson Model A tractor with road bands.

a hole in the rim so that it dug into the soil surface. As the wheel continued turning, bringing the spud clear of the soil, the action of the cam brought it back inside the rim again. The movement of the spuds was designed to scrape off clinging soil in order to maintain maximum traction, and the cam system was standard equipment on the Ideal after about 1917.

Another feature of the Ideal tractor, which was built in Birmingham for about ten years from 1912, was an adjustment to vary the length of spud protruding through the rim. The maximum length available for the most difficult soil conditions was 4 in., but this could be reduced until nothing protruded. With the spuds adjusted to remain within the rim, there was no need to fit special road rims as the outside surface of the wheels remained smooth.

Some American manufacturers also offered various forms of self-cleaning spuds and lugs but these, like the system used on the Ideal, were usually a complicated and expensive way to deal with mud.

(*Following page*) Vickers Aussie tractor with self-cleaning rear wheels.

Vickers used a less complicated method for controlling mud on the rear wheels of their Aussie tractor announced in 1925.

VICKERS
VICKERS LTD
MADE IN ENGLAND

Solid rubber tyres on a 1921 chain driven Lanz Bulldog.

Although the tractor was made in England, it was built mainly for the Australian market, which accounts for the curious name.

Each of the driving wheels was made in three sections with a gap between each of the sections. Steel bars were attached to the engine sump and passed through the gaps between the wheel sections. If mud started to build up on the wheel rims the bars would help to knock it off.

The 30 hp tractors remained in production for about five years, and most were exported to Australia. Equipment to clear mud from tractor wheels seems a curious feature to sell in Australia, but the idea had originally been developed on an Australian farm.

Tractors used mainly for stationary operation and for road work were often supplied with solid rubber tyres. An example of this was the Avery Farm and City tractor, which ran on wooden pegs for farm work, but for which rubber tyres were available for city roads (see page 58).

Inflatable tyres were also tried during the late 1920s, but they were usually high pressure tyres designed mainly for highway use, and they lacked the flexibility and deep tread pattern necessary for field work.

1. A 1903 Ivel from the Hunday Museum in Northumberland.

2. This 30hp Marshall tractor was exported to Australia in about 1908 and brought back to England 75 years later for restoration.

3. Holt tracklayer made in 1910 with a four cylinder petrol engine and a single steering wheel at the front.

4. International Harvester 12–25 Mogul tractor made in about 1914.

5. The 10–20 was the most popular model in the International Titan range.

6. Driver's view of the 10–20 Titan.

7. The Waterloo Boy Model N became part of the John Deere product line when the Waterloo Gasoline Engine Co was taken over by Deere in 1918.

8. View from the Waterloo Boy driving seat.

9. Alldays & Onions 30hp tractor equipped with leaf springs at the rear and a coil spring over the front axle.

10. The boiler shaped structure on the Walsh & Clark Victoria cable ploughing tractor is the fuel tank.

11. Twin cylinder engine mounted on top of the Victoria fuel tank.

12. The three-wheel drive Glasgow tractor was made in Scotland between 1919 and 1925.

13. When General Motors followed Ford into the tractor market in 1917 they bought the Samson company and developed the Model M tractor.

14. The Austin was Britain's answer to the Fordson, powered by a four cylinder engine designed for the 25hp Austin car.

15. This is a 1917 version of the Moline Universal with a horizontally opposed two cylinder engine.

16. The Moline Universal engine.

17. Fiat's first production tractor in 1919 was the 702 with a 25hp engine.

18. Rumely 16–30 Oil Pull tractor with an oil-cooled engine.

19. A special lever was supplied with the Rumely to make it easier to turn the flywheel when starting the engine.

20. The first of the Bulldog models from Lanz with a 12hp semi-diesel engine.

21A. The single cylinder engine of the 1921 Bulldog.

21B. Allrad Bulldog made in 1923 with articulated steering.

22. Hanomag built this 50hp Z50 tracklayer in 1924.

23. An early 1920s version of the Model F Fordson, the most successful tractor model ever built.

24. Rear mudguard modification on a 1927 Fordson.

25. The Hurlimann 1K10A series tractor built in 1929 with a single cylinder semi-diesel engine developing 12hp.

26. Deutz began building the MTZ 220 tractor in 1931 with a twin cylinder engine developing up to 30hp.

27. Fordson paint colour was changed from grey to blue when production was transferred from Cork to Dagenham in 1932.

28. The Massey-Harris General Purpose with four-wheel drive through equal size wheels.

29. Lanz Eil Bulldogs were designed for transport work.

30. Ferguson Model A tractor made by David Brown with Harry Ferguson's hydraulic implement control system.

31. Vierzon was the leading French manufacturer of semi-diesel tractors during the 1930s.

32. The new look Massey-Harris Pacemaker arrived in 1938 to replace the angular lines of the previous model.

33. An early version of the Bristol tracklayer made in 1934 and powered by a 10hp Austin car engine.

34. Allis-Chalmers Model M tracklayer powered by a four cylinder distillate engine developing up to 35hp.

35. Fordson with Roadless tracks and a winch, built in 1941 for use by the RAF.

36. International Harvester W4 built in 1941 with 23hp maximum output.

37. Caterpillar D4 exported to Britain in 1943.

38. The Allis-Chalmers Model B remained in production for 20 years from 1937. This is a 1943 example.

39. International Harvester Model H, one of the most popular tractors in the Farmall series.

40. Minneapolis-Moline tractors were given more up-to-date styling in 1939 and the Universal Z series is an example.

41. This 45hp tracklayer version of the Lanz Bulldog was built in 1945.

42. The County Full Track or CFT tracklayer was based on the Fordson E27N, and this 1950 example is equipped with a Perkins P6 diesel engine.

43. British built International Harvester BMD with a Perkins diesel engine.

44. The implement lift mechanism on the David Brown 2D tool carrier was powered by air.

45. The Doe Dual Drive or Triple-D tractor with two engines and one set of controls.

46. Porsche tractor built in 1962 with a single cylinder diesel engine.

47. The John Deere 4020 tractor was available with an optional single lever powershift transmission.

Allis-Chalmers Model U manufactured in about 1939.

The breakthrough came in 1932 when the Allis-Chalmers company tested a set of aircraft tyres on one of their Model U tractors. The tyres were designed to operate at only 15 psi, a relatively low inflation pressure, and tests showed that this helped to increase traction because the tyre casing remained flexible enough to mould itself around uneven surfaces and stones.

The tyres tested on the Model U tractor provided an important breakthrough in tractor development. The Model U became the first tractor that could work efficiently in the field and then travel on the highway. Leading companies such as Firestone and Goodyear in America and Dunlop in Britain quickly developed special tyres for tractors, but many farmers were concerned about durability and punctures and preferred to use iron wheels.

Allis-Chalmers decided to use speed to promote the 35 hp Model U and its tyres, and during 1933 they organised one of the most ambitious and entertaining publicity campaigns the tractor industry has seen.

The company formed a team of specially modified high speed tractors to race at the principle state fairs, with well known racing car drivers such as Ab Jenkins and Barney Oldfield at the wheel for some of the races. They also drove the tractors in a series of speed record attempts, and by the end of the campaign Ab Jenkins had pushed the world speed record for tractors to 67 mph. The record attempt on the Utah salt flats was officially observed by the American Automobile Association. Presumably the record still stands.

(Following page) **Allis-Chalmers Model U racing tractor on inflatable tyres.**

Another development that made an important contribution to tractor efficiency was four-wheel drive. Putting power through four driving wheels increases traction by 10 per cent or more in difficult soil conditions, especially with equal diameter front and rear wheels, but there were some mechanical problems to be overcome before four-wheel drive tractors became popular.

Four-wheel drive was available on steam traction engines before the first tractor had been built. The Rubicon traction engine, built in the mid-1880s by Wood, Taber and Morse of Eaton, New York, was equipped with a gear drive to the front and rear wheels to provide increased wheel grip on muddy roads and tracks. However, demand remained small and most customers chose conventional two-wheel drive models.

Several American tractor companies were experimenting with four-wheel drive by about 1910, but few farmers were interested in the idea and tractors, such as the Olmstead Four-Wheel Pull made from about 1912 in Great Falls, Montana and the 1919 Samson Iron Horse four-wheel drive, sold in small numbers.

One of the mechanical problems was that of providing a reasonably sharp turning angle with powered front wheels, and for many years farmers who bought four-wheel drive tractors had to put up with poor manoeuvrability. The Pavesi company in Italy overcame the problem by taking the steering function out of the front axle on their P4 tractor, which was available during

Pavesi P4 tractor with four-wheel drive and articulated steering.

the 1920s. This allowed a reasonably sharp turning angle while providing four large diameter driving wheels.

The P4 tractor was equipped with an articulated steering system, with front and rear sections joined at a hinge point in the centre. The tractor was powered by a water-cooled engine with overhead valves and force-feed lubrication.

In 1927 the Armstrong-Siddeley car company in England announced an agreement to manufacture the P4 tractor. Although the two companies signed the agreement, according to a press report at the time, there is little evidence that any of the tractors were produced at the Armstrong-Siddeley factory.

The bend-in-the-middle steering arrangement allowed the designers to use solid axles front and rear, and the same arrangement was used in 1923 by Lanz for the four-wheel drive or Allrad version of their Bulldog tractor. The unique feature of the Allrad Bulldog was the diameter of the wheels, with the front wheels bigger than those at the rear.

Although sales of four-wheel drive tractors remained small, interest slowly increased during the 1920s and 1930s. More companies moved into the market, including Massey-Harris with their General Purpose model.

The General Purpose arrived in 1930 with four-wheel drive through equal size wheels. Massey-Harris had previously bought their designs from other companies such as Parrett and the J I Case Plow Works, and the General Purpose was the first tractor designed and manufactured by Massey-Harris. Production continued until 1936, and the tractor was probably the most popular four-wheel drive model at that time.

After the end of the Second World War in 1945, the four-wheel drive market began to expand more rapidly, particularly in Britain and other European countries. Specialist companies moved into the market, including Bray, County, Doe, Muir Hill and Roadless, and these provided the first opportunity for most farmers in Britain to try the advantages of four-wheel drive.

The most successful of the specialists was County, which began building the CFT tracklayer based on a Fordson E27N base unit in 1948. Their first four-wheel drive model was the Four Drive in 1954, with four equal size wheels. It had no steering wheel, but used the same skid-steer system with levers as the CFT tracklayer.

Super Four and Super Six tractors followed the Four Drive in 1961. They were equipped with twin drive shafts and steered front wheels, and were the first of a succession of four-wheel drive County tractors that remained in production until 1990.

The Doe Triple-D arrived in the 1950s, also based on Fordson components. The idea of joining two tractor units together to provide extra pulling power was developed by George Pryor,

County Super 4 tractor with four-wheel drive.

an Essex farmer with a large acreage of heavy arable land. He wanted more pulling power for ploughing, but as there was no suitable four-wheel drive tractor available he decided to build his own.

He used two of the new look Fordson Major tractors with the front axle and wheels removed from both. The front of one of the tractors was attached to the rear of the other with a hinge point to provide articulated steering, and the controls of both were linked to allow the two engines and transmissions to be controlled from the driving seat at the rear.

The result was an 80 hp tractor with four-wheel drive and articulated steering. It attracted considerable interest and Ernest Doe, a leading Ford tractor distributor, began making the Triple-D on a commercial scale. Production started in 1957 and reached more than 300 when the last Doe tractor was built in 1966.

High horsepower four-wheel drive tractors were built on a similar basis by specialist manufacturers in North America. The two leading independent companies were Versatile in Canada and Steiger based in the United States.

The first Steiger tractor, like the first Triple-D, was built on a farm where extra pulling power was needed. The Steiger brothers

Doe Triple-D four-wheel drive tractor with two engines.

farmed 4000 acres near Red Lake Falls, Minnesota and they began building their first tractor in 1957.

They wanted a big four-wheel drive machine, and the tractor they built in the farm workshop used a 130 hp engine, four equal size driving wheels and articulated steering.

Interest from other farmers encouraged them to build more tractors, and they formed a company offering three models with up to 318 hp. As the business expanded a new company was formed in 1969, and production was transferred to Fargo, North Dakota. Steiger had close links with Ford for several years, and Steiger tractors were sold in Ford colours. This arrangement ended when Case IH bought the Steiger company, and Ford then took over Versatile to provide an alternative range of high horsepower four-wheel drive tractors with articulated steering.

The only way to improve the pulling efficiency of a tractor with four-wheel drive is to use crawler tracks. Although their principal attraction is reduced wheelslip, tracks also help to spread weight over a large area to minimise soil compaction, and this is a considerable advantage on less stable soils which are easily damaged by excess ground pressure.

The biggest disadvantage of traditional steel crawler tracks is the lack of mobility on the road, and they also have a reputation for being slower and noisier than wheels.

Crawler tracks, like four-wheel drive, first appeared on steam powered equipment. There were several British and American attempts to design suitable tracks during the nineteenth century,

A 110hp Best steam engine working with a combine harvester in about 1900.

but costs were high and reliability was generally poor. Tracklayers continued to be more expensive than ordinary models, but design improvements—mainly in the United States—brought increased reliability, and the market for crawler tractors began to expand in the early years of the present century.

The outstanding names in crawler tractor history are Best, Holt and Caterpillar. Daniel Best of San Leandro and Benjamin Holt from Stockton, both in California, were rivals in the combine harvester business during the 1890s. They also built the steam traction engines to pull the combines, and this made them aware of the problems of operating heavy machinery on the soft soil in some of the grain growing areas. Both used giant wheels to spread the load, and one of Best's traction engines built in 1900 was supplied with wood covered driving wheels 9 ft high and 15 ft wide. Holt supplied an even bigger machine with triple wheels on both sides, and each wheel was 6 ft wide.

Holt built his first experimental tracklayer in 1904. It was designed for cultivating particularly soft areas of land, with the driving wheels of a steam traction engine replaced by a set of tracks equipped with wooden shoes. The idea proved successful and more steamers with crawler tracks were built. In 1906 the Holt company built their first crawler tractor powered by one of their own gasoline engines.

In 1908 Best agreed to sell his company to Holt, his main competitor, but two years later the Best name reappeared in the tractor industry when Daniel Best's son started the C C Best Gas Tractor Co. Both companies were building crawler tractors, and rivalry between the two families was soon re-established.

Meanwhile the Holt company had opened a new factory in Peoria, Illinois in 1909, and during the following year the Caterpillar trade name was formally registered. Holt had already been using the Caterpillar name for about five years, and this was the name chosen for the new company formed in 1925 when Holt and Best agreed to join forces again in a merger.

Caterpillar has remained the leading tracklayer company, and has introduced most of the significant developments in crawler tractor design. The first diesel powered tracklayer was a Caterpillar, and it was also the first production tractor in America with diesel power (see page 81). During the 1980s Caterpillar announced the Challenger, the first production tracklayer with reinforced rubber tracks for travelling on the land or on a road.

One of the developments that helped to focus attention on crawler tracks was the tank designed for the British Army in the First World War. Tanks provided an effective demonstration of track mobility in difficult conditions, and the British Army also used modified Holt tractors for transport work.

(Facing page)
(Top)
Holt 70–120 tracklayer made in Peoria in about 1920.

(Bottom)
Caterpillar Sixty photographed in 1930.

(*Left*)
Hornsby-Akroyd tractor with Roberts track system in 1904.

(*Below*)
Strait's tractor with up-turned rear tracks in a 1914 photograph.

The British Army had been interested in the idea of using tracked vehicles for transport work for many years, and one of the candidates was a tracklaying version of the Hornsby-Akroyd tractor from Grantham. David Roberts, the Hornsby-Akroyd chief engineer, designed a new type of track and this was tested on one of the company's tractors.

Although the War Office tested the tractor, the big army contracts failed to materialise. Some reports suggest the track patents were sold to Holt in America, but there is little evidence that Holt put Hornsby-type tracks into production.

When war was imminent, the Army looked to America for a track system for the secret Landship or tank project. Tractors imported for evaluation included the Bullock Creeping Grip from Chicago and the Strait's Tractor made at Appleton, Wisconsin.

The pre-war Bullock tractor had tracks at the rear, wheels at the front for steering, and the driver sat in the middle, behind the engine but in front of the fuel tanks. A more conventional 12–20 tractor arrived in 1916, with no wheels but with steering clutches to control the tracks.

The layout of the Straits Tractor was also unconventional. There were two tracks at the rear and one much smaller track at

A 1919 Renault GP tractor based on a wartime tank.

the front for steering. As the front track was not visible from the driving seat, a pointer was fitted to the top of the front steering unit to show which way the track was facing.

William Strait, who held the patents to the tractor, had designed the main tracks with a sharply tilted rear profile. This made it easier to reverse the tractor over obstacles, and it is one of the features that seems to have impressed the War Office when the tractor was demonstrated. The upward tilted track shape appeared on the first army tanks that helped to break the stalemate of mud and trench warfare.

Tanks were developed from farm tractors during the war, but the process was reversed when the war ended. The French Army used a small, highly successful tank made by Renault during the war, and this was the basis for the first Renault tractor, the GP tracklayer, announced in 1919.

Clayton and Shuttleworth crawler tractor made in 1920.

Fowler VF tracklayer with a single cylinder diesel engine.

The GP was powered by a four cylinder petrol engine which developed 30 hp, and had the typical Renault arrangement of a radiator behind the engine instead of at the front. It was also equipped with an unusual tiller arrangement for steering the tractor. The GP remained in production for about a year, and was replaced by a modified version called the H1. The HO version with wheels instead of tracks followed later.

Several British companies moved into the crawler tractor market after the war including Clayton and Shuttleworth, Blackstone and Sentinel. There were also others, like Rushton, with tracklayer versions of their standard models. In the 1950s the County CFT became Britain's first commercially successful agricultural crawler tractor, competing against the Fowler VF tracklayer. The VF was effectively the crawler version of the Marshall tractor, and it remained in production until 1957, still using the single cylinder diesel engine which first appeared in 1930.

Howard Rotavator, previously a machinery specialist, entered the crawler tractor market after taking over the Fowler company in Leeds. Fowler was later sold to Marshall of Gainsborough, but

(*Facing page*)
(*Above*)
Bogmaster version of the Howard Platypus tracklayer made in the 1950s.

(*Below*)
Ransomes MG5 mini-sized crawler tractor.

Howard remained in the tracklayer market making the Platypus range of tractors at a factory in Basildon, Essex from 1950 until 1958. Most models were powered by Perkins engines, and several versions were available including the Bogmaster models with extra-wide tracks for working in soft soil conditions. The Bogmaster could be purchased with the Bogwaggon, a trailer with similar width tracks which were powered from the tractor p-t-o.

Britain's top selling crawler tractor was the Ransomes MG. The letters MG stand for Market Garden, and the mini sized tractor was not designed for agricultural work. It first appeared in 1936 as the MG2 powered by a 6 hp air-cooled petrol engine, and production continued through various versions for 30 years until the sales total reached 15,000. Most were used in nurseries and smallholdings, but the MG was also popular in France for vineyard work and some were sent to Tanzania for use in salt pans.

The final version of the Ransomes tracklayer was the MG40 powered by a 10 hp diesel engine.

Other European countries also used crawler tractors, particularly Italy where maximum traction and stability were needed for deep ploughing and for hillside work. In Germany there were tracked versions of the Lanz Bulldog, and the Hanomag company built the Type Z50 tractor in 1926, a mighty tracklayer with a four cylinder petrol engine of almost 9 litres capacity.

Although Best, Holt and Caterpillar occupied most of the

Bates Steel Mule Model C in a 1917 photograph.

John Deere Lindeman tracklayer conversion for orchard work.

limelight in the American tracklayer market, there have been plenty of competitors.

The tracklayer with the most memorable name and the most distinctive appearance was probably the Bates Steel Mule. The Bates Machine and Tractor Co of Joliet, Illinois built a whole series of Steel Mule tractors, but the one many people remember is the Model C. This had a number of distinctive features including the rocket shaped cladding over the engine and fuel tank, and the tricycle layout with two wheels at the front and a single rear track. Another unusual feature of the Model C was the telescopic steering column allowing the steering wheel to be moved up to eight feet to the rear. This, with other extendable controls, allowed the tractor to be driven from a seat on a trailed implement.

Companies with less eccentric but more commercially successful crawler tractors included International Harvester, starting with their TracTractor series in the 1930s.

The Monarch Tractor Co of Watertown, Wisconsin built a successful series of tracklayers from about 1913, some of which had

Cletrac Model R made in 1917.

appropriate names such as Lightfoot and Neverslip. When Allis-Chalmers decided to move into the tracklayer market, they bought the Monarch factory and product range in 1928.

John Deere concentrated on wheels rather than tracks until 1949 when the first of the MC tracklayers was built. Between 1940 and 1947 John Deere enthusiasts who preferred tracks were offered the John Deere Lindeman Crawler. This was produced by the Lindeman brothers of Yakima, Washington for orchard work, but the tractor was popular in other areas for jobs as varied as logging and general cultivations.

The Lindeman tractor was based on the John Deere B0 skid unit with a two cylinder engine developing about 14 hp. Tracks were available in a choice of widths from 10 to 14 in.

The principal competitor for Best, Holt and Caterpillar tractors was the Cletrac range from Cleveland, Ohio.

The company was established in 1916 as the Cleveland Motor Plow Co, and its first production tractor was the Model R with the neat sales slogan, 'Geared to the Ground'. It was a small tractor, powered at first by a Weidley four cylinder engine providing 20 hp at the belt pulley and a rather modest 12 hp at the drawbar. A Cleveland engine was used on later versions.

During the 1920s and 1930s the company, known as the Cleveland Tractor Co and using the Cletrac trade name, expanded its product range to cover virtually all sectors of the market. Their biggest tractor, announced in 1927, was called the 100, with a 100 hp rated output from a six cylinder Wisconsin engine. The first diesel powered tractor, the Model 40 powered by a Hercules six cylinder engine, arrived in 1934.

The Cleveland company was taken over by the Oliver tractor company in 1944, and Oliver became part of the White Motors group in 1960. White stopped making crawler tractors during the 1960s.

3

Tractors for Transport

SOME tractors spend 60 per cent or more of their working time doing transport jobs or travelling, and this is why JCB decided to develop their high speed tractor for the 1990s.

The importance of transport work has been recognised by manufacturers who have designed special models since the early days of tractor development, and some of the first British built tractors were designed as load carriers.

The first vehicle built by H P Saunderson was a load carrier for agricultural and general transport work. It was completed in time to compete in the Royal Agricultural Society trial of 'Self-Moving Vehicles' organised in conjunction with the 1898 Royal Show. The trial included 13 miles of public roads, but the Saunderson

Saunderson transport vehicle made in 1898.

entry failed to complete the trial because of an engine fault. The single cylinder engine with a 6½ in. bore and 10 in. stroke had been designed to run on gas but was converted to petrol for the RASE trial.

Another vehicle which failed to complete the course was the Roots and Venables entry. During the road test the driver lost control of the steering after one of the front wheels was damaged in an accident with a car. In 1898 there were only a few hundred cars in Britain and the chances of being hit by one of them must have been extremely small, but the impact was sufficient to put the Roots and Venables in a ditch and out of the contest.

The Saunderson vehicle used a chain and sprocket drive with two forward speeds. The brief details still available do not include any mention of a reverse gear. The front axle was equipped with leaf springs.

The price of the Saunderson vehicle in 1898 was £250. It was included in the company's catalogues for several years, but it is doubtful if any were sold. The original prototype version that had broken down during the RASE trials was taken back to the Saunderson works near Bedford, and according to someone who had worked for the company, it was lying derelict there in 1906, and was eventually scrapped.

The Saunderson company was founded in 1890 after H P Saunderson came back to England from Canada. While in Canada he had made contact with the Massey and Harris companies, and the farm machinery business he established included the agency for equipment made by the recently formed Massey-Harris company.

Tractor production gradually expanded after about 1903, and in 1905 the company had 17 employees. Most of the tractors were three-wheelers designed for general farm work, with a detachable platform over the rear wheels for load carrying.

This arrangement was featured in a demonstration in 1906 when a Saunderson tractor was used to pull a binder in a field of wheat. The binder was then removed; the load platform was fitted and it was used to transport the sheaves to a nearby threshing machine. The next job was using the tractor pulley belt to drive the thresher, and the newly threshed grain was then ground in a mill also powered by the belt pulley. A baker made bread with some of the flour while the tractor, with the load platform removed, ploughed and cultivated the freshly cleared stubble and sowed a new crop of wheat.

The whole process of using one tractor to harvest wheat, turn the grain into bread and drill another crop of wheat was completed in five hours.

Four-wheel tractors in the range included the Model L, listed

Saunderson Model L transport tractor available in 1910.

in the Saunderson catalogue for 1910. This was powered by a single cylinder, air-cooled petrol/paraffin engine which Saunderson described as 6 to 8 hp. The rear platform could be tipped, and was also removable. Like other models in the catalogue, the paint finish was French grey with red wheels 'unless otherwise preferred'.

Sales remained small until Saunderson stopped making the transport type models and began selling ordinary farm tractors. The Universal series, and particularly the Model G which was available throughout most of the wartime sales boom, helped to make the Saunderson company Britain's biggest tractor manufacturer for just a few years, and some Universals were also assembled in France where they were sold as Scemia tractors.

The success was short-lived. The demand for tractors slumped in the early 1920s, but competition increased as more companies moved into the market. In 1924 H P Saunderson sold his company to the Manchester-based Crossley engineering group, and one of their ideas for boosting the sales figures was to market a

Petter's Patent Agricultural Tractor appeared at the 1903 Royal Show.

transport version of the 25 hp Saunderson tractor for towing road trailers with a maximum gross weight of 10 tons.

Other British manufacturers attracted by the transport tractor idea included the Petter company in Yeovil and the Coventry-based Daimler company. Petter's Patent Agricultural Tractor was announced at the 1903 Royal Show. It was equipped with a platform mounted on a four-wheel chassis with leaf springs, and the driving seat was positioned right at the front of the tractor to leave a large unobstructed load space. The water-cooled, single cylinder paraffin engine was equipped with hot tube ignition and developed 12 hp.

According to the Petter company, the tractor was suitable for field work, including ploughing, and could be used as a stationary power unit with the belt pulley. It was also a transport vehicle for carrying or towing produce from the farm to a railway station.

The gearbox on the Petter provided reverse and two forward gears, with a maximum travel speed of 4.5 mph. This was about as fast as other tractors at that time, and it was presumably a satisfactory travel speed for farmers who had previously worked at the slow pace of horses or a steam traction engine.

Daimler provided three forward gears for their new 30 hp tractor in 1911, with a top speed of 7 mph in third. The four cylinder engine featured the patented sleeve valve design that

Daimler announced their new 30hp tractor in 1911.

had helped to make Daimler cars among the quietest available. It could operate on petrol only or petrol and paraffin, and the power was transmitted through a leather faced cone clutch.

Equipment provided in the standard specification included an automatic lubrication system operated by a plunger pump, and an unusual safety device which automatically locked the transmission gears in neutral to prevent the tractor being moved accidentally while the belt pulley was working. The load container behind the cab was also standard equipment, but the tractor was promoted mainly for ploughing and cultivation work, and transport appears to have been a secondary function.

American manufacturers were generally less interested in the transport potential for their tractors, but the Avery company was an exception. Their Farm and City tractor arrived on the market in 1909 with a truck body for load carrying and a front mounted belt pulley to power stationary equipment.

Avery was one of the first companies to realise that a transport vehicle should have a reasonably fast travel speed, and their new tractor could reach 15 mph with the four cylinder engine running at more than 1000 rpm and producing a maximum output of 36 hp. The front axle was equipped on both sides with a leaf spring suspension to give a smoother ride, and there were two seats with the driver on the right hand side.

The Avery tractor/truck was built for several years from 1909.

Customers were offered a choice of wheels. Tractors to be used mainly on the road were fitted with solid rubber tyres, but for working on the land there were special steel wheels with small indentations or cups around the circumference. A hardwood peg was hammered into each of the cups, and this was supposed to minimise wheelslip. The sales figures proved to be disappointing and production ceased after about four years.

French manufacturers produced small numbers of transport tractors during the 1920s and 1930s. The best known of these were the Latil tractors with four-wheel drive through equal size wheels. They were popular for jobs such as timber haulage, but they were also suitable for agricultural work and were equipped with a pulley for driving stationary equipment.

The Latil KTL tractor, entered for the 1930 World Trials held near Oxford, developed 28.2 hp in the belt test at the trials and 19.8 hp at the drawbar from a four-cylinder engine operating at up to 1900 rpm. The high specification included a gearbox with two reverse ratios and six forward gears with up to 17 mph maximum travel speed. Stopping power was provided by a foot operated transmission brake plus brakes on all four wheels.

During a haulage test in the Oxford trials the Latil handled a gross load of 10 tons without problems. Demand for the tractor was limited by the price, which was £655 in Britain at a time when ordinary tractors with more power than the Latil were available for less than £250.

Citroen-Kegresse half-track vehicles were based on the front end of a truck, with tracks at the rear and space behind the driver's seat for carrying a small load. They were first available in

(*Facing page*) Citroen-Kegresse transport tractor (*above*) photographed at a 1922 ploughing demonstration and (*below*) loaded up for military use.

V.6
V.6
V.6
2272W1

about 1921, and were still on the market ten years later, but the sales total was probably small.

The Citroen-Kegresse was more successful as a cross-country transport vehicle than as a farm tractor, and they were sometimes in the news when used for geographical expeditions.

The Citroen-Kegresse transport tractor made a successful appearance in the 1930 World Tractor Trials. Maximum output from the four cylinder petrol engine was 16.4 hp on the belt and 13.1 hp in the drawbar tests. Maximum speed in sixth gear was 14.6 mph and the gross load in the towing tests was 4 tons.

The biggest market so far for transport tractors has been in Germany where most of the leading manufacturers have built one or more models designed for haulage work.

Deutz made tractors for moving heavy military equipment during the First World War, and these were followed in 1919 by a tractor designed for haulage and general farm work.

The power unit was a 40 hp petrol engine, and the design included a sprung front axle. The tractor could be supplied with a winch, and this made it particularly suitable for forestry work. It was too big and expensive for most farmers, and the sales volume remained small. Deutz achieved more success when they began making smaller farm tractors with a simple semi-diesel engine.

The first really successful transport tractor was the 55 hp HR-9 version of the Lanz Eil Bulldog, which was available from 1937 to 1944 and achieved a production total of more than 2400.

An earlier Eil Bulldog had proved there was a substantial market for a transport tractor. The previous model was powered by a single cylinder semi-diesel or hot bulb engine developing up to 45 hp. Eil means fast, and the first Eil Bulldog had a 12.5 mph top speed. This was faster than most conventional farm tractors in the 1930s, but Lanz realised that a further increase in the top gear performance would make longer journeys possible and expand the market.

For the new HR-9 Eil Bulldog, Lanz used the same type of engine with one cylinder of 10 litres cubic capacity. They raised the engine speed from 650 to 750 rpm, increased the power output to 55 hp, and provided a five speed gearbox to boost the maximum speed to 20 mph. The fuel tank capacity was raised to 200 litres or 44 gallons. Full road lighting was provided plus direction indicators and electric windscreen wipers, and the front axle was fitted with a transverse leaf spring suspension.

A cab was standard equipment, and customers could choose either a solidly built version or a cabriolet style cab with a canvas hood which could be folded down in fine weather. Inside the cab there was a padded bench seat for two, with the driver positioned on the right instead of the usual left-hand drive arrangement.

The semi-diesel engine could be started on petrol using an

Deutz started making a heavy haulage tractor in 1919.

electric motor operated by a foot switch in the cab and powered by batteries carried on the running boards. If this failed the usual hand starting process could be used, with the steering wheel used to turn over the engine.

Eil Bulldog customers included timber merchants, builders and transport contractors as well as farmers, and sales peaked at more than 500 in 1939. Having proved that the market for a transport tractor existed, it was surprising that Lanz decided not to continue the Eil series after the War.

The company which replaced Lanz in the transport tractor market was also German. The Mercedes-Benz Unimog was developed mainly for the farming industry, and it achieved considerable success as a versatile power unit for a wide range of jobs including load carrying, ploughing and operating p-t-o powered machinery. Later versions of the Unimog have become even more popular for military and industrial use.

The original Unimog was designed by Albert Friedrich who used a 25 hp Mercedes-Benz diesel engine, four-wheel drive through equal size wheels, and four-wheel braking. He also

An early version of the Unimog working with front mounted equipment

provided a power take-off and a hydraulically operated rear linkage, a belt pulley and front p-t-o and linkage options. Newer versions have considerably more power, and the maximum speed has been increased from the original 31 mph to more than 50 mph.

Unimog production averaged 6000 per year throughout most of the period from the 1960s through the 1980s. Military purchases account for 40 per cent of the production total, with about 12 per cent sold to farmers and contractors. Although agriculture is now only a relatively small proportion of the sales total, the Unimog remains one of the most important and innovative developments in power farming.

While later versions of the Unimog have become more of a truck than a tractor, the Mercedes-Benz MB-trac was designed specifically as an agricultural tractor with a well developed transport capability. The mid-mounted cab allows only a small load carrying space over the rear wheels, but the MB-trac has a 25 mph top speed and four-wheel braking for haulage work on the road.

4

Providing the Power

POWER farming started at the end of the eighteenth century with steam. It was the steam engine that offered farmers their first real alternative to animal power, and stationary engines were used in Britain to power threshing machines and feed processing equipment almost 100 years before the first tractors were available.

Stationary steam engines were followed by portable engines pulled by horses, and later by self-propelled traction and cable ploughing engines. By the end of the nineteenth century, when

Steam traction engines like this Nichols and Shepard provided the first real alternative to animal power.

the first tractors were being built, steam power was well established on big farms and estates in the main arable farming areas.

Tractors made slow progress at first, mainly because of poor reliability at a time when steam engines had the advantage of more than 100 years of commercial development. Many of the problems were caused by primitive ignition, and the first fuel systems were particularly sensitive to working in dusty or damp conditions.

Improvements in engine reliability helped to open up the market, and this was an important factor in making tractor power more widely accepted.

Another factor was the power-to-weight ratio of the internal combustion engine which allowed smaller, cheaper tractors to outperform steam powered traction engines. The big disadvantage of the traditional steam engine was its size and weight. An engine with enough power to pull a plough or drive a threshing machine has to be big and heavy, and this can be a disadvantage when a traction engine has to be driven across a field while the soil is moist and easily compacted.

Tractors also suffered from weight problems at first, as the results of the Winnipeg Agricultural Motor Competitions showed. These competitions, which were held annually in the period leading up to the First World War, appear to have been well organised and they attracted most of the leading American and Canadian manufacturers plus the Marshall company from England.

The results provided an interesting performance comparison at a time when steam engines and tractors were competing for sales in the rapidly expanding Canadian market.

In the 1910 competition the Gold Medal for steam engines of 60 hp or less was awarded to a Case traction engine weighing 7.8 tons and developing 60 hp in the maximum load brake test, equal to 7.69 hp per ton. The Gold Medal award for tractors of 30 hp and above went to a Gas Traction from Minnesota, which also weighed 7.8 tons and produced 6.9 hp per ton. The most powerful tractor in the 1910 competition was the Rumely Oil Pull, weighing almost 12 tons and developing a decidedly modest 4.13 hp per ton from a petrol/paraffin or gasoline/kerosene engine.

While traction engine manufacturers continued with their heavy, traditional design, tractor development was making rapid progress. The first batch of Nebraska tests was held in 1920, and the results show a big improvement in the power-to-weight figures for some of the tractors entered. An International Harvester Junior weighing 1.63 tons managed 11.36 hp per ton, the figure for the Fordson was 15.83 hp per ton, but the Oil Pull E, a 26,000 lb survivor from the heavyweight era of the American tractor industry, produced only 6.51 hp per ton.

The 30–60 Rumely Oil Pull tractor weighed 26,000 lb (12 tons).

The figure for a modern four-wheel drive Case IH Magnum 7130 tractor weighing 17,540 lb is 26.18 hp per ton.

Other steam engine limitations were also highlighted in the detailed results recorded by the judges at the Winnipeg competitions. Each of the 18 entrants in 1910 had to complete a two-hour engine test on a brake to measure the power output, fuel consumption and the amount of water evaporated from the tractor cooling systems or lost from the steam engine boilers.

Three of the tractors, a Kinnard Haines and one from both the American and Canadian Gas Traction factories, achieved zero water loss, and easily the highest loss figure was 15 gallons from a twin cylinder International Harvester tractor. The six steam engines in the competition averaged 475.6 gallons during the six hours, with a 120 hp Rumely steamer topping the thirst league with 623 gallons.

Internal combustion engines also scored when the weight of fuel used during the two-hour test was compared. The fuel used by the tractors ranged from 21.5 lb of petrol or gas for the single cylinder International to 107.25 lb of paraffin or kerosene used by the 50 hp Rumely Oil Pull tractor. Coal consumption for the six steamers averaged 644.6 lb, including 861 lb used by the 120 hp Rumely traction engine.

Petrol engines were almost standard equipment during the first 20 years of tractor development, and the 50 hp Rumely was the only tractor of the 12 competing at Winnipeg in 1910 to have a petrol/paraffin engine instead of just petrol. Ten years later only 12 of the 65 tractors completing the Nebraska test programme in

Rumely tractors, like this 16–30 Oil Pull, were among the first to use paraffin or kerosene.

1920 used petrol and the others were equipped to burn paraffin.

The Rumely company of LaPorte, Indiana, later known as the Advance-Rumely Thresher Co, helped pioneer the use of cheaper fuels such as paraffin or kerosene instead of petrol, but they were not the first with this type of engine. The Hornsby-Akroyd tractor from England was equipped with a single cylinder hot bulb petrol/paraffin engine in 1896.

The Rumely engine had two big horizontal cylinders, the typical design used by American tractor companies at the time, but it also had two unusual features. The cooling system was filled with oil instead of water to take heat from the cylinder block, and the engine was equipped with a patented injection system to add water to the fuel in the combustion chamber.

Oil cooling was used for two reasons. It avoided frost damage problems in cold weather—an important factor in the Midwest and Canada where many Rumely Oil Pull tractors were used—and it also allowed the engine to run at a higher temperature to help burn the paraffin more efficiently.

The water injection system, designed to give smoother operation when burning paraffin, was originally developed for stationary engines by John Secor, an American engineer. He joined the Rumely company in 1908 and helped to produce the first Rumely tractor engine, using the special Secor-Higgins carburettor he had designed and patented with William Higgins, a manager at the Rumely factory.

The cooling process for Rumely Oil Pull tractors took place in a large rectangular tower over the front wheels. Hot oil from the cylinder block passed through vertical plates inside the cooling tower. The plates, similar to central heating radiators, were spaced to allow air to pass between them to take away some of the heat, with the exhaust outlet placed just above the radiator plates and pointing upwards to encourage an air flow from the base of the tower.

Oil Pull tractors, with their distinctive cooling towers, were popular in America and Canada. They were ruggedly built and, by contemporary standards, reliable. Smaller, lighter Rumely tractors were built, but the Oil Pulls with their slow speed, horizontal paraffin engines continued to dominate the Advance-Rumely range during the 1920s and production continued until 1931 when the company was taken over by Allis-Chalmers.

Petrol continued to lose its popularity as more American manufacturers equipped their tractors with paraffin burning engines during the 1920s, and this type of power unit continued to dominate the tractor industry for the next 30 years or so.

Four cylinder engines powered most of the mid-range tractors, but John Deere continued to fit a horizontal twin in almost all their tractors between 1918 and 1960. During this period the distinctive 'two-lunger' engine was almost as much a John Deere symbol as the deer emblem and the green and yellow paint, and loyal customers objected when John Deere began building tractors with a more modern engine design.

The John Deere two cylinder engine was mounted with the top of the cylinders facing towards the front end of the tractor and the horizontal crankshaft towards the rear, with the belt pulley attached to the end of the crankshaft. The engine was simple but successful, with fewer moving parts than the more conventional four cylinder layout, and it was still in use when the Model R, the first John Deere diesel tractor, was launched in 1949.

Engine development in the British tractor industry was influenced by the Fordson, International Harvester and other American makes during the 1920s; Austin, Glasgow, Rushton and Vickers-Aussie tractors were all powered by four cylinder engines which reflected American thinking. In fact the Rushton engine was so closely influenced by the Fordson design that some parts were interchangeable.

There were also some manufacturers who ventured away from the standard design, and these included Peter Brotherhood who made the 30 hp Peterbro tractor at Peterborough from 1920 until the early 1930s.

The Peterbro power unit was a four cylinder engine based on a Ricardo design which aimed at overcoming the problem

(*Following page*) The Model AA tractor was one of a long line of John Deeres with a two cylinder engine.

JOHN DEERE
JOHN DEERE
GOOD YEAR

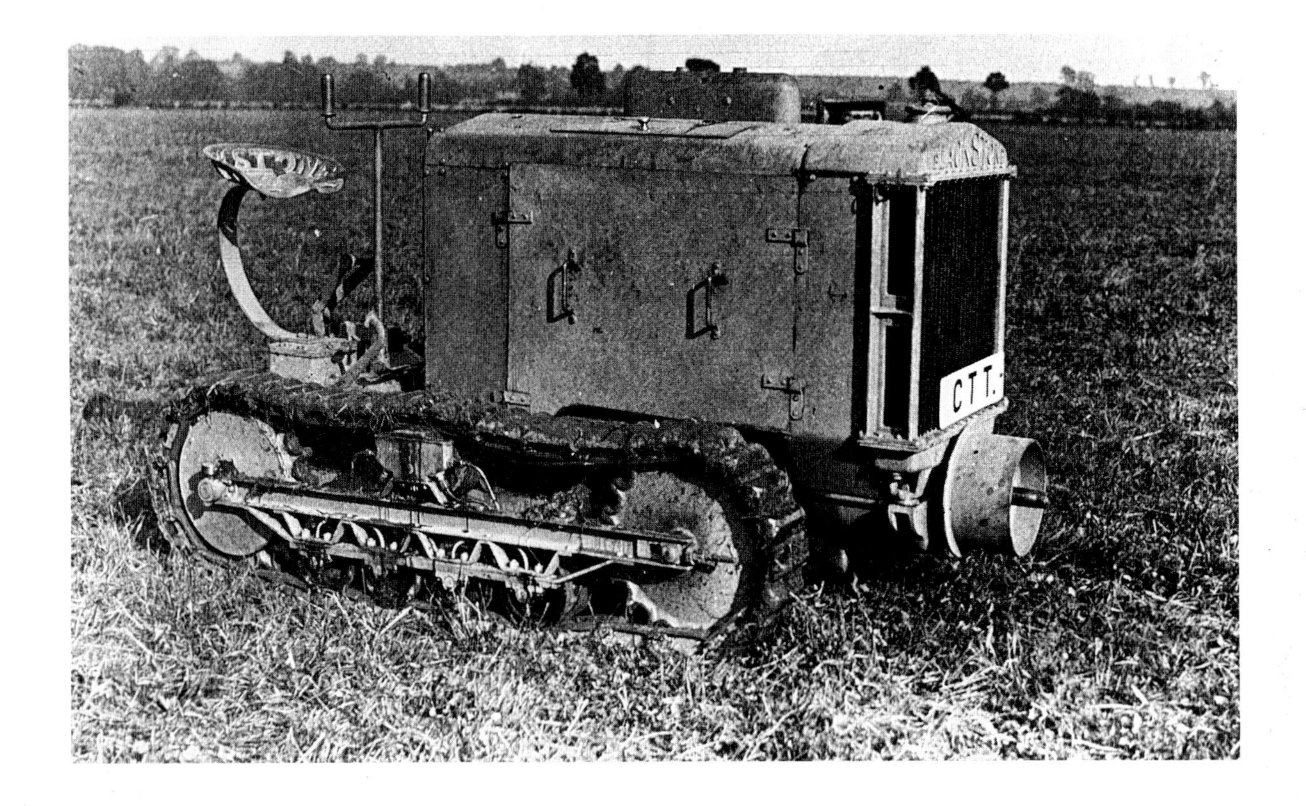

The engine on this Blackstone crawler tractor started from cold on paraffin.

of unburned paraffin being forced past the pistons to dilute the sump oil. This is a characteristic of paraffin engines which most manufacturers accepted, but the Peterbro engine was designed with a cross-head layout and with a series of tiny bleed holes drilled through the cylinder walls to divert waste paraffin away from the sump. Extra production costs helped make the Peterbro more expensive than most of its competitors, but in spite of this it was popular in Australia and New Zealand.

Another British variation on the standard design was the three cylinder engine used in the Blackstone tractor of 1919. This was unusual because it was designed to start from cold on paraffin.

A compressed air system was provided on the Blackstone to turn the engine over and avoid using the starting handle, and special injection equipment was used to squirt paraffin into the combustion chamber to help the cold engine ignite the fuel. This system appears to have worked well, but it was complicated and expensive at a time when farmers preferred simple, low cost tractors.

There was also a demand for petrol/paraffin tractors in other European countries, and Fiat and Renault both moved into the market in about 1919 using this type of power unit.

Tractors with semi-diesel or hot bulb engines were also popular on the continent, though they made little impact elsewhere. They first appeared just before the beginning of the First World War

(*Facing page*)
(*Above*)
A 1929 Renault PE tractor with the radiator behind the four cylinder engine.

(*Below*)
Close up of the 1929 Renault PE engine with the radiator behind the engine.

One of the first Lanz Bulldog tractors with a single cylinder semi-diesel engine.

A 1924 HSCS tractor from Hungary with a 14hp semi-diesel engine.

in 1914, and they were built in substantial numbers between 1919 and the 1950s. The best known manufacturer was the Lanz company in Germany, now part of the John Deere organisation, but many others specialised in this type of engine including Deutz, also in Germany, Hofherr-Schrantz-Clayton-Shuttleworth or HSCS in Hungary, Landini in Italy, Munktells in Sweden and Vierzon, the leading French semi-diesel manufacturer.

Semi-diesels are compression-ignition engines which rely on a hot spot inside the combustion chamber to help ignite the fuel

(*Above*)
The blowlamp for starting this 1922 Lanz Bulldog engine is in a special holder below the cylinder head.

(*Right*)
Single cylinder semi-diesel engine of 14,300 cc capacity on a 1932 Landini Super.

charge. When the engine is cold a blowlamp is used to produce the hot spot and allow the tractor to start, and this procedure may take 20 minutes or more on a cold morning. Once the engine is running the hot spot is maintained by the heat released during the combustion process.

Most of the semi-diesels were single cylinder two-stroke engines, positioned so that the cylinder head was right at the front of the tractor to make the heating process more convenient.

Compared to the four cylinder engines used by most American and British manufacturers between the two world wars, the semi-diesels had some disadvantages. They were slow speed engines with a relatively low power output and modest torque back-up, and they were not as smooth as a multi-cylinder power unit.

A power output comparison between a semi-diesel and a four cylinder petrol/paraffin engine was included in my book *Massey-Ferguson Tractors*, first published in 1987. The engines compared were the 14,300 cc single cylinder semi-diesel designed for the 40 hp Landini tractor announced in 1932, and the petrol/paraffin four cylinder engine, based on a Wallis design, which powered the Massey-Harris 20–30 tractor, also produced during the early 1930s. The rated power output per litre of engine capacity was 2.8 hp for the semi-diesel and 5.3 hp for the Massey-Harris engine. The comparable figure for a modern diesel engine with a turbocharger is about 25 hp per litre.

Semi-diesels also had some important attractions, and for many farmers these outweighed the disadvantages. The single cylinder with no electrical ignition equipment was unbeatable for mechanical simplicity, and this helped ensure reliability and ease of maintenance at a time when experienced service fitters were almost unknown in many rural areas.

Many of the engines were also well designed and solidly built to give a long working life, and durability was encouraged by the slow working speed and modest power output.

Another attraction of the semi-diesel is its ability to burn almost any type of liquid fuel. This can help to reduce running costs, and it has also proved useful during periods when fuel is difficult to obtain, as a semi-diesel tractor can continue working on waste products such as tar oil or sump oil diluted with paraffin.

Semi-diesel tractors were still being built in small numbers during the 1950s, and the last Landini tractor with this type of power unit left the factory in Italy in 1961.

Lanz played a leading role in the development of the semi-diesel tractor engine, and the Lanz Bulldog wheeled and tracklaying models were the most successful tractors of this type during the 1920s and 1930s. Production started in 1921 when the first of the Bulldog tractors left the Mannheim factory with a 12 hp engine designed by Fritz Huber.

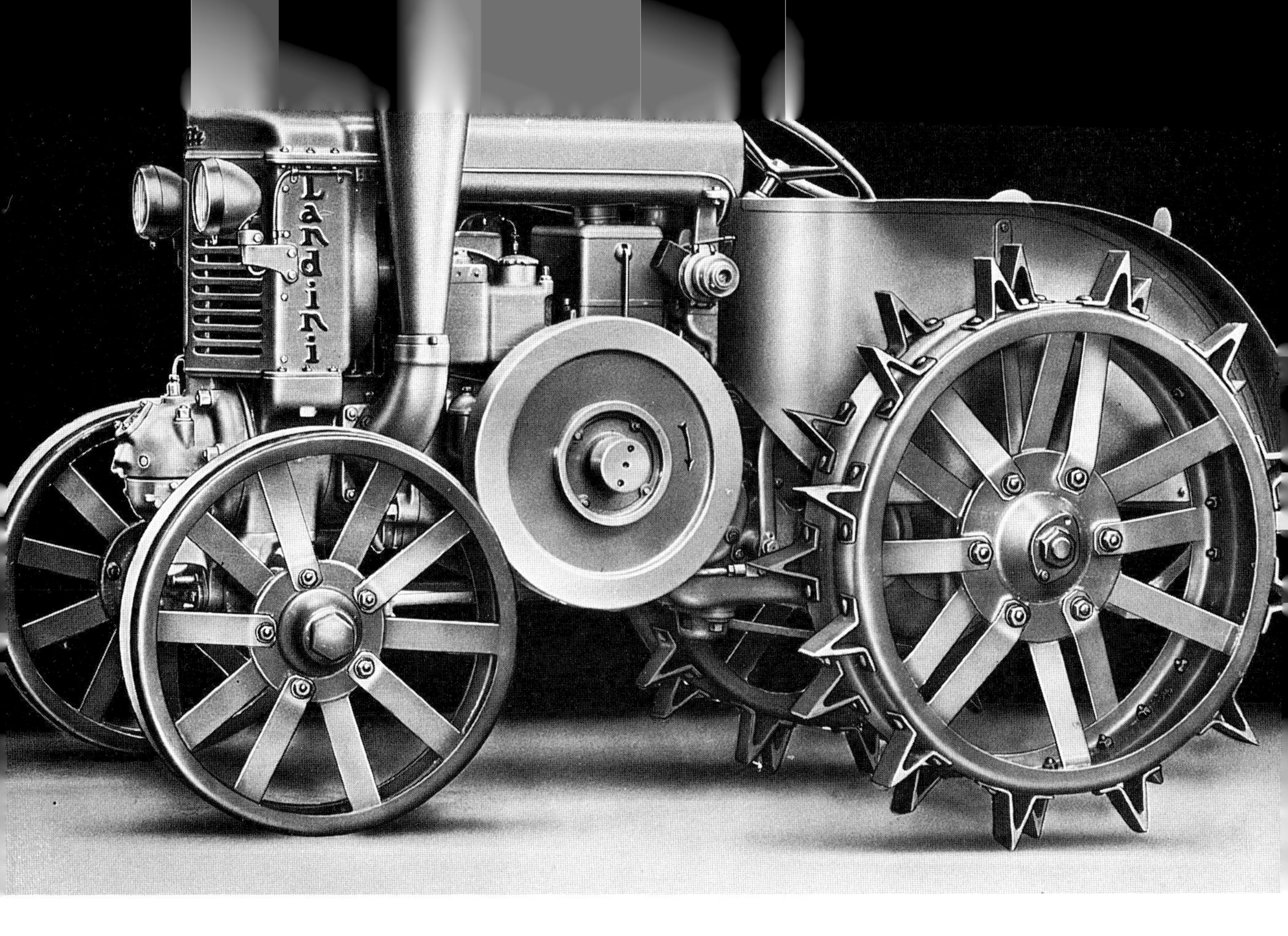

The Landini Velite semi-diesel tractor was available from 1935.

The engine worked equally well with anti-clockwise or clockwise rotation. This meant that the transmission needed forward gears only as the tractor could be driven backwards by reversing the engine direction. There were wooden blocks which the driver could press against the rear wheel rims to stop the tractor, and this was probably sufficient to cope with the 2.6 mph top speed on the road.

A new model in the Lanz Bulldog series in 1923 had an advanced specification with articulated or bend-in-the-middle steering and four-wheel drive with front and rear wheels of almost equal diameter—in fact the front wheels were slightly bigger than those at the rear. Further development in the 1930s included the Eil Bulldog or express tractor for transport work (see chapter 2).

There were six models in the Bulldog 06 series, which arrived in the 1930s and was phased out in the early 1950s. Some 06 models were equipped with a six-speed gearbox at a time when most of their competitors offered three or four ratios.

Diesel powered tractors arrived during the early 1920s, and they were also a German development. The first Benz tractors ran on petrol, including a 50 hp tractor built in 1919 with a four cylinder power unit of 8.1 litres capacity. It was still in production during the early 1920s when Benz began experimenting with diesel engines to power trucks and tractors.

Diesel engine production started on a small scale at the Benz factory in Mannheim during 1921 or 1922. The twin cylinder engine was probably used to power a truck at first, but in 1923 the same type of engine was available in the Benz-Sendling motor plough. This was a curious machine with two wheels at the front and one large driving wheel at the rear. It had been

Cylinder head of a 1936 Lanz Bulldog engine rated at 55hp.

made since 1919 with a petrol engine, but the 1923 version was almost certainly the first commercially produced tractor with a diesel engine. A more conventional four-wheel diesel tractor was available until about 1931.

In 1926 the Benz and Daimler companies merged to form Daimler-Benz, manufacturing the Mercedes-Benz vehicle range which later included the MB-trac, one of the most important developments in modern tractor design.

Manufacturers were quick to appreciate the value of the semi-diesel engine and the number of companies offering this type of power unit increased rapidly during the 1920s, but the Benz diesel brought a more cautious response.

This was partly, perhaps, because the diesel represented a bigger, more ambitious advance in technology than the simple semi-diesel. Another factor may have been the extra production cost of a diesel engine, which helped to force up the price of the tractor at a time when the farming industry in much of Europe was under financial pressure.

Francesco Cassani of Italy was another of the diesel pioneers. He learned most of his engineering skills helping to make and repair implements for local farmers in his father's workshop, and this was the start of a career which included designing the first high speed marine diesels for the Italian navy and diesel engines for aircraft.

The 1923 Benz-Sendling, the first diesel tractor.

One of the first Cassani 40hp diesel tractors built in the late 1920s.

Cassani designed and built his first diesel tractor in 1927 when he was 21 years old. The prototype, with a vertical, twin cylinder water-cooled engine, was followed by a small production batch of tractors which are variously described as 40 and 45 hp. The engine on these was equipped with a starting mechanism operated by compressed air from a high pressure tank. The air supply in the tank was replenished while the engine was running to build up the pressure ready for restarting.

Tractor production ended during the 1930s, but Cassani formed a new company in 1942 to manufacture SAME tractors, and the Lamborghini and Hurlimann tractor companies were later added to the SAME Group.

British manufacturers were also involved in diesel tractor development during the late 1920s, and when the World Tractor Trials were held near Oxford in 1930, three of the five diesel entries were British made. The others were a single cylinder Mercedes-Benz and a tractor entered under the McLaren name but based on a four-wheel version of the Benz-Sendling with a twin cylinder diesel engine.

In commercial terms the most significant of the three British diesel tractors in the 1930 trials was the Marshall powered by a single cylinder, two-stroke engine. It was based on a prototype completed in 1929, and it was the first in a long series of single cylinder Marshall and Field Marshall tractors that remained in production until 1957.

The cylinder head of the Marshall engine was pre-heated with

(Facing page)
(Above)
Industrial version of the Marshall 18/30 diesel tractor made in 1932.

(Below)
A 1930 picture of a Blackstone tractor with a four cylinder diesel engine.

BLACKSTONE

a blowlamp to make starting easier, and this system was also used to start the early Benz and Mercedes-Benz diesels. Marshall tractor engines were equipped with cartridge starting in 1945 when the Series 1 Field Marshall was announced.

The 1930 Marshall engine had an 8 in. cylinder bore and 10.5 in. stroke, and the compression ratio was 16.5 to 1. The maximum power output in the belt tests at the World Trials was 29.1 hp, with 21.4 hp in the drawbar tests, exceeding the manufacturer's 24 and 16 hp power ratings.

The other British diesel entries were the Blackstone and the Aveling and Porter, both made by the Agricultural and General Engineers (AGE) group.

AGE was formed when a large group of British farm machinery manufacturers decided to pool their resources to create a company powerful enough to challenge the increasing success of big North American firms such as Case, Deere, International Harvester and Massey-Harris. It was an imaginative idea, but in commercial terms it was a dismal failure and when the world trials were taking place in 1930 the AGE group and its member companies were on the brink of a financial disaster.

The engine in the Blackstone tractor was a four cylinder diesel, and there was also a small petrol engine for starting the main power unit. The rated output on the pulley belt was 26 hp, but this was easily exceeded in the 1930 tests with a 37.7 hp maximum.

Aveling and Porter provided the engine for the second of the AGE entries, but it was mounted in a tractor made by Garrett, another of the companies in the group. This was also a four-stroke, four cylinder diesel, though an electric starter motor was provided. The rated output was 22 hp on the drawbar and 38 hp on the belt, and the maximum output figures in the 1930 tests were 30 and 42.4 hp.

The 1930 trials provided the first opportunity to compare the performance of many of the world's leading diesel, petrol, paraffin and semi-diesel tractors under strictly controlled test conditions, and in the fuel consumption tests the diesel tractors provided clear evidence of their superior efficiency.

Fuel efficiency is measured in horsepower hours per gallon of fuel, and a high figure means more of the energy in each gallon of fuel is available for useful work. Measurements are taken for drawbar pull and power on the belt pulley, in both cases with the tractor engine working at its rated power output. Drawbar performance is affected by wheelslip and the efficiency of the transmission, and the output on the belt is the best indication of engine efficiency.

The five diesel tractors in the test averaged 16.31 hp hours/ gallon, compared to 12.55 for the four semi-diesels, 9.46 for the ten petrol tractors, and 9.84 for the nine tractors competing in the paraffin section. As diesel was very much cheaper than either petrol or paraffin in 1930, the fuel cost for diesel powered tractors was significantly less than for those with a spark ignition engine.

Although diesel tractors scored on fuel costs, they were more expensive to buy. The lowest priced diesel at £315 was the single cylinder Marshall with a 16/24 power rating, the 20/26 hp Blackstone cost £500 and the Mercedes-Benz 14/20 price was £360. Paraffin powered alternatives tested in the 1930 trials included a 20/30 Massey-Harris for £300, a 17/27 Case Model C for £248 and the 11/15 hp Austin cost £210.

Diesel tractors made little impact in the European market during the 1930s, apart from the modestly successful Marshall. Progress was also slow in the United States where the most important exception was Caterpillar which started a diesel engine development project in 1929. A prototype tractor was tested in 1930 and Caterpillar became the first American manufacturer to market a diesel tractor in 1931.

By 1937 most models in the Caterpillar range were offered with a diesel engine option. Caterpillar diesels were supplied with a small, water-cooled petrol engine to provide the power to start the main diesel unit, and the small engine also helped to warm up the water in the cooling system to encourage the big diesel engine to start more easily.

One of the most popular models in the Caterpillar range was the R2 with a choice of petrol and paraffin engines and the D2

Mercedes-Benz tractor with a single cylinder diesel engine entered in the 1930 World Trials.

diesel version announced in 1938. All three engines were based on the same cylinder block design with the same cylinder dimensions, and they all produced about 29 hp. They were all tested at Nebraska in 1939 and the results provide further evidence of the extra fuel efficiency of a diesel tractor. In the belt tests, the figures for horsepower hours per US gallon of fuel used were 13.24 for the diesel powered D2, 8.62 for the petrol version of the R2 and 8.99 per gallon when burning paraffin.

Caterpillar led the way in the development of diesel tractors in America during the 1930s, but diesel options were soon available from other leading tracklayer manufacturers. International Harvester announced their TD-40 diesel powered TracTractor in 1932 and the Cletrac 80 diesel was added to the Cleveland Tractor Co range in 1933. There was less interest in diesel power for wheeled tractors, but the WD-40 was included in the International Harvester range from 1935, powered by the four cylinder engine already used in the TD-40 tracklayer.

The Waukesha-Hesselman engine was an unusual variation on the standard diesel design. It was available in the United States during the 1930s, and made a brief commercial appearance on the Bates Steel Mule 35 tractor in 1934. The engine had a combination of spark ignition with fuel injection, and the aim was to allow cheaper fuel to be used, with the spark ignition system added to overcome the starting problems associated with some of the early diesels. Spark plugs also avoided the high compression ratio of a conventional diesel engine.

Diesel tractors proved their superior fuel efficiency during the 1930s, but there was little response from customers until the late 1940s.

The breakthrough came when some of the leading tractor manufacturers began offering diesel versions of popular mid-range models. Design improvements had produced engines

(*Above*)
Caterpillar D2 diesel tractor.

(*Facing page*)
(*Left*)
Injection equipment for a Caterpillar four cylinder diesel engine made in 1940.

(*Far left*)
Perkins L4 diesel engine in a 1948 International BMD Super.

(*Below*)
Six cylinder Leyland diesel engine in the Marshall MP6 announced in 1954.

which were easier to start, and bigger production volumes helped reduce manufacturing costs to make diesel tractor prices more attractive. It was also a period when British farmers were encouraged to increase food production, and there were financial incentives to encourage increased investment in new equipment.

Perkins engines provided diesel power for several British

Turner Yeoman of England tractor powered by a 34hp V-4 diesel engine.

tractor manufacturers. The P6 engine, first developed in 1937, was available as an option for the E27N Fordson Major and as standard equipment in the Massey-Harris 744D in 1958. Other Perkins powered tractors included a diesel version of the Nuffield Universal announced in February 1950 with a P4 engine, the Ferguson TE series, and the P3 engine offered as an option for the Allis-Chalmers Model B from 1954.

David Brown developed a diesel version of their paraffin engine for the Cropmaster in 1949, Marshall announced the new MP6 tractor in 1954 using a six cylinder Leyland engine and the Minneapolis Moline UDS tractor was available in Britain in 1949 with the choice of Dorman or Meadows diesel engines.

One of the most unconventional diesels in the late 1940s was the 34 hp V-4 engine in the Turner 'Yeoman of England' tractor announced at the 1949 Royal Show. The cylinders were at a 68 degree angle, and the cylinder heads bulged out on both sides of the engine compartment, giving the Turner a distinctive

Fordson Super Major diesel engine made in 1960.

appearance. The Turner engine was designed by Freeman Sanders, who had previously developed diesel engines for the Fowler company.

The launch of the Fordson New Major at the 1951 Smithfield Show in London was also an important diesel development. The tractor was available in petrol, paraffin and diesel versions, all with overhead valve engines. The diesel engine earned a reputation for easy starting, smooth operation and good reliability, and it marked another important stage in the development of the diesel as the standard power unit for farm tractors.

5

Engine Alternatives

PETROL, paraffin and diesel engines have powered most of the tractors built during the last 100 years or so, and there have also been large numbers of multi-fuel hot bulb or semi-diesel tractors. Other types of engines and fuels have been used on an experimental basis, and sometimes commercially, but so far with no significant impact on tractor development.

Attempts to use electricity for ploughing are almost as old as the petrol tractor. One of the earliest reports came from Germany in 1894 when the Zimmermann company of Halle demonstrated their electric ploughing equipment.

Zimmermann showed two different systems, both based on a balance plough of the type developed in England by John Fowler and Ransomes to work with a steam powered cable system. Instead of a steam engine and winding drum on the headland, Zimmermann used an electric motor attached to the plough. The motor drove two spur gears which engaged in the links of a chain stretched across the field, and the plough could be driven backwards and forwards along the chain by reversing the current.

Both ends of the chain had to be firmly anchored, and each time the plough reached a headland the anchors and the plough had to be moved a short distance on to unworked land. The electricity supply was taken to the motor through a cable from the headland, with a series of supports to avoid dragging the cable across the soil. These supports also had to be moved at regular intervals as the ploughing progressed.

One of the 1894 prototypes was based on a plough which turned four furrows each way, and this was equipped with a 16 hp electric motor linked to a mains supply brought to the field through a cable. The other version of the Zimmermann system was powered by a 10 hp motor with a two-furrow plough, and a steam engine on the field headland drove a generator to provide the electricity supply.

Moving the anchors, the plough, the heavy chain and the electric cable supports, and also maintaining a supply of steam for the generator for the smaller plough, made the Zimmermann system complicated and laborious.

Brutshke electric tractor for cable ploughing.

The Brutshke tractor was a more successful German attempt to use electricity for ploughing. It was demonstrated on land near a sugar beet processing factory where there was surplus electricity available from the steam powered generating plant to power the tractor. Overhead wires were provided to carry the electricity supply from the factory to the field.

The tractor was used with a cable ploughing system. It consisted of a simple four-wheel chassis carrying the electric motor, enough electric cable to link the tractor to the nearest power supply point, plus a winding drum for the cable to the plough.

The 220 volt electric motor powered a winding drum with a steel cable to pull a plough or cultivator to and fro across the field while the tractor stayed on the headland. The motor was also used to bring the tractor to the field and to move it along the headland as the plough progressed across the field.

Brutshke tractors were in use on some of the bigger farms in sugar producing areas in about 1902, and operating costs were said to be lower than a steam powered cable system.

A mechanisation specialist writing in a British magazine in 1903 suggested that the benefits of an electric ploughing system could justify the cost of installing a steam powered generator on farms with a big arable acreage.

Ploughing with electricity avoided the need for coal and water for steam engines in the field, but there were also some limitations. Overhead power lines to bring the power supply to the electric motors would have been costly to install, and the trailing cable linking the motor to the power line must have been difficult to protect from abrasion.

The Brutshke system fell from favour because petrol tractors provided a more convenient alternative, but experiments with electricity for field work continued.

**(*Facing page*)
Late 1940s pictures of electric tractors working in Russia.**

A Frenchman, M. Felix Prat, was another pioneer of ploughing by electric power. In 1895 a disused water mill on his estate at Enquibaud, Tarn was adapted to drive a generator to provide a power supply for buildings on his estate. He also fitted an electric motor on a small tractor unit equipped with a capstan and a cable drum.

The cable drum was used to pull a plough backwards and forwards across the field, and a rope on the capstan pulled the tractor along the headland as each set of furrows was turned.

The stream that powered the water mill sometimes dried up during the summer, and M. Prat realised that this would cut off the power supply to all the electric motors installed on his estate. To provide a more reliable water supply he built an elaborate series of reservoirs which were linked to the stream by a canal system.

In spite of the problems, experiments with electricity for ploughing continued, particularly in Russia where cable systems were developed with limited success during the 1930s.

Russian engineers switched their research from cable systems to direct ploughing with tractors powered by electric motors, and this proved to be a more promising approach. Both track-laying and wheeled tractors were used during the 1940s, and the Russian engineers provided an elaborate network of power lines to bring the current to the fields on some of the big collective farms.

The current passed through a mobile transformer on the headland and was taken to the electric tractors through a cable which was wound out from a drum as the tractor moved away from the transformer, and wound in again while working towards the power source.

Information published by the Russian Agricultural Ministry in 1951 claimed all the practical problems had been overcome and more farms were being cultivated by electric tractors. Between 1945 and 1949 the number of tractors had increased by 300 per cent, the Russians claimed, and further expansion was confidently predicted.

Russian mechanisation specialists used electric tractors on big farms with large fields, a situation where the lack of flexibility of a

Ransomes tracklayer powered by electricity.

Allis-Chalmers fuel cell tractor.

tractor plugged in to a power supply is likely to be a considerable limitation. A British research team chose the opposite approach, developing an electric tractor suitable for small acreages. The Electrical Research Association started work on the project in 1949 at Reading, Berkshire, using a mini-sized Ransomes MG crawler tractor equipped with a 9 hp electric motor instead of the usual petrol engine.

The power supply to the tractor came from the top of a 35 ft pylon through a wire plugged in to a special connector which could turn through 360 degrees, allowing the tractor a 110 ft working radius around the base of the pylon. A system of counterweights attached to the pylon kept the power line under tension while the tractor worked.

A report published in 1953 included enthusiastic comments about reduced noise levels, easier starting and cheaper engine maintenance, but the research team also admitted that the extra cost of the supply cable and the pylon might be difficult to justify on a small acreage.

American attempts to harness electricity for field work have concentrated on self-contained tractor units, and one of these was the Allis-Chalmers fuel cell tractor built in 1959.

A fuel cell is a form of battery which converts chemical energy

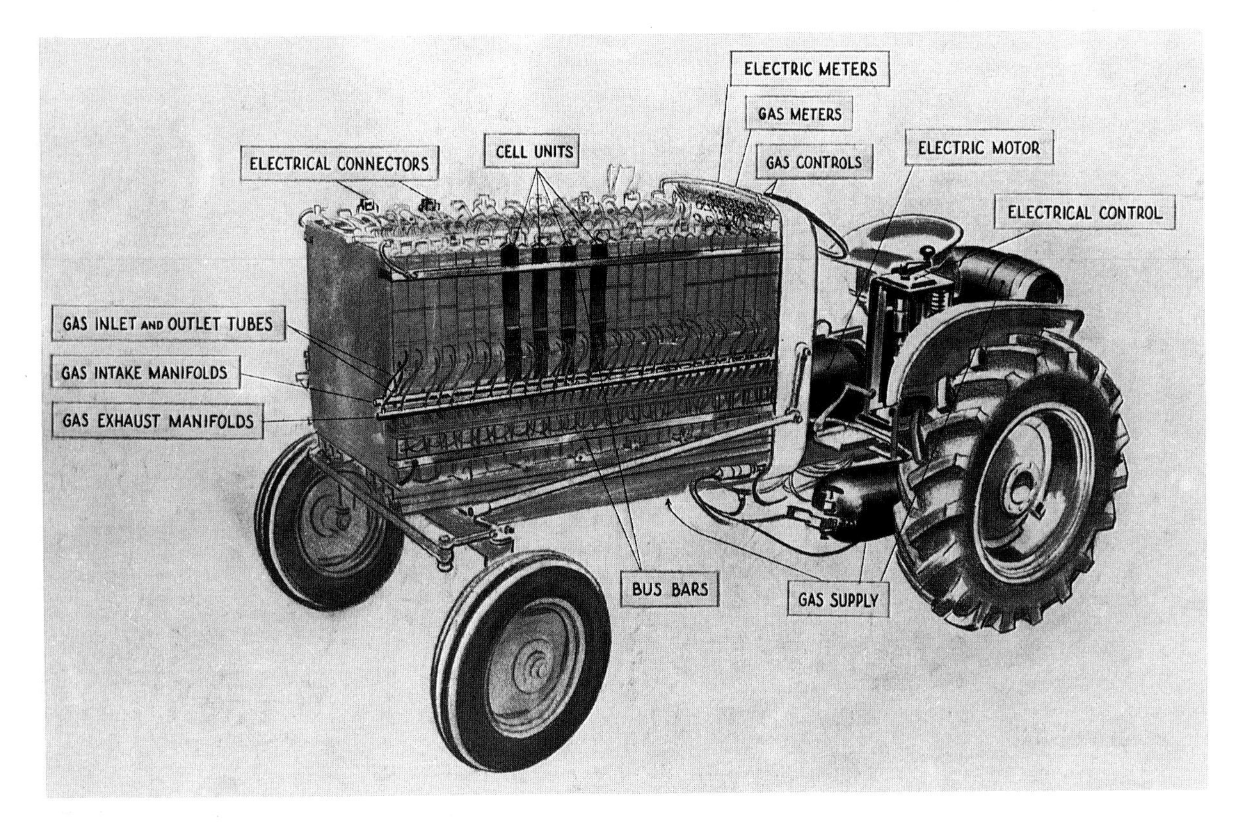

Diagram of the Allis-Chalmers fuel cell tractor.

into electricity. Chemicals in the form of gases, solids or liquids are fed into the fuel cell causing a reaction which produces an electric current. Unlike an ordinary battery, the fuel cell is not able to store the electricity it produces.

The basic theory of the power cell was first described by a British scientist in 1839, but since then the idea had been regarded as a scientific curiosity. Allis-Chalmers was one of several companies taking a new look at fuel cell technology in the 1950s, attracted by the theoretically high energy efficiency of the chemical to electricity process.

When diesel fuel is burned in an engine, 70 per cent or more of the energy in the fuel is wasted. In a fuel cell the energy losses could be 10 per cent or less.

The first Allis-Chalmers tractor experiments were carried out in a converted D-12, and these were followed by a specially built research tractor with a block of 1008 small fuel cells. Chemicals supplied to the cells were a mixture of propane and other gases which were carried on the tractor in pressurised cylinders. Varying the supply of gas instantly altered the amount of electricity produced, and the driver could use the supply tap to regulate the forward speed. The electricity was produced as a direct current to power a 20 hp motor which drove the rear wheels of the tractor, and changing the polarity of the current to the motor reversed the tractor.

Although the energy efficiency of the cells is relatively high, losses occur as the motor converts the electrical energy into mechanical power to drive the rear wheels, and the overall efficiency of the experimental fuel cell tractor was not much higher than a diesel. Disadvantages included the large amount of space that was occupied by the banks of fuel cells, which were also heavy and which gave the tractor a poor power to weight ratio.

Fuel cell tractor experiments have so far proved to be a dead end, and this also seems to be the fate of the Choremaster experimental tractor developed by a research team working in the Agricultural Engineering Department at South Dakota State University.

The tractor was powered by the current from two 32-cell battery blocks providing 43.5 kWh of electric power to operate two direct current or DC motors. One of the motors drove the pump for the tractor hydraulics and steering circuits plus the power take-off, and the other powered the four-wheel drive tractor through a three range gearbox and hydrostatic transmission.

Work on the Choremaster project started in 1983. The research team had accepted that battery power has too many limitations for jobs such as ploughing, and the experimental tractor was designed mainly for yard work.

Limitations of the experimental tractor were the fact that it carried two tonnes of batteries plus two electric motors. The batteries had to be recharged from a main supply, and this took up to eight hours to prepare the batteries for about six hours of work.

Potential advantages claimed for battery powered tractors include low noise levels and virtually pollution free working. The energy efficiency was claimed to be superior to that of a diesel tractor performing similar tasks, but this probably does not include energy needed to generate the electricity used to recharge the tractor batteries.

Further progress with battery powered yard tractors may be possible if there are advances in battery performance, but meanwhile the South Dakota State University team has been developing a battery powered skid steer loader for farm work.

Steam engine technology made rapid progress during the early 1900s with smaller, lighter power units being developed for use in cars. American manufacturers such as Stanley and White made steamers which overcame most of the disadvantages of traditional steam power, and for a decade or more they provided effective competition for petrol powered cars.

Most of the big companies making agricultural steam engines in the United States either ignored the new technology or waited too long before using it. There were also some, such as J I Case and Rumely, who were wisely moving out of the traction engine

Bryan Harvester steam tractor made in Peru, Indiana.

business and into tractor production, but they were the exceptions, and most of their competitors simply continued to make steam traction and portable engines along traditional lines for a dwindling market.

The technology that was waiting to be exploited included liquid fuel burners that could be adjusted by turning a tap, thus making bulky coal supplies, the stoker's shovel and clearing the ashes all unnecessary. The new style boiler was tubular with just a small volume of water, and it took five minutes or less to reach working pressure from a cold start.

Another development adopted by steam car manufacturers long before it arrived in the agricultural market was a really efficient condenser. This recirculated water from the used steam to reduce the need for topping up.

The new steam technology was available on the American tractor market for several years from 1922 when the Bryan Harvester Co of Peru, Indiana made the 20 hp Model A, a tractor they said was 'good old steam power modernised'.

The tubular boiler, a radiator type condenser and the two cylinders of the power unit were so compact that they were all housed in the engine compartment of what looked like just an ordinary tractor. The Bryan was about the same size as other 20 hp tractors, and the weight of 5500 lb was less than that of some of its rivals with an internal combustion engine.

Paraffin burners were used to heat the water in a tubular boiler. The boiler was designed to work at up to 600 psi, and there was a safety device which automatically turned off the fuel supply to the burners when the maximum working pressure was reached.

The two horizontal cylinders had a 5 in. stroke and 4 in. bore and developed the rated power output at 220 rpm. They were equipped with piston type valves operated by a Stephenson linkage, with force-feed lubrication provided by Madison-Kipp equipment. The 60 gallon water tank was big enough for a day's work, and the fuel tank held 30 gallons.

The Model A was advertised as 'The World's First Light Steam Tractor', and the sales literature stressed the benefits of steam power for jobs such as threshing. But in spite of this, demand proved to be disappointing.

International Harvester was also developing a tractor in the early 1920s with state-of-the-art steam technology. Their commitment to the project is shown by the surviving photographs of a series of prototype tractors built between 1920 and 1923. The final version appears to be carefully finished and may have been a pre-production prototype, but at that stage development work came to an end. The project may have been shelved to allow more resources for the Farmall tractor launch, or the budget may have been axed because the company was facing intense sales pressure from the Fordson.

The IH tractors, like the Bryan steamer, used paraffin to heat water in a tubular boiler. The engine was a vee-twin, and it was linked to a condenser to conserve water.

Most of the British manufacturers, who had done much of the original development work for the traditional agricultural steam engine, reacted slowly to the challenge of the internal combustion engine. Ransomes was an exception, demonstrating a tractor in 1903, and at the Marshall factory they hedged their bets from about 1907 by making tractors running on paraffin and coal-fired traction engines.

Garrett made their first break from traditional steam engine design in 1917 when they built the first Suffolk Punch. It was a cross between a steam tractor and a traction engine, and was designed for ploughing as well as road haulage and pulley belt work. To emphasise the new concept Garrett called it an agrimotor.

One of the novel design features was placing the driver right at the front of the Suffolk Punch. This ensured excellent forward visibility, but the view to the rear was poor.

The horizontal boiler was behind the driver, with the coal burning firebox towards the front and the chimney at the rear. The specification included a superheater to raise the steam temperature to 315 degrees C (600 degrees F).

INTERNATIONAL

JET 1, the Rover car powered by a gas turbine.

Ackermann steering was used to ease the driver's job, and this also allowed an arched front axle to increase the ground clearance. Both axles were fully sprung to give a better ride over rough roads.

The Suffolk Punch agrimotor arrived in 1917, the same year as the Fordson. Suffolk Punch production totalled eight—one of which has survived—and when production stopped in 1928, sales of the Model F Fordson had reached almost 750,000.

Another unsuccessful attempt to challenge petrol/paraffin and diesel engines for powering tractors was the gas turbine. During the late 1950s and early 1960s many people believed it was the power unit for the future, and assumed turbines would soon replace petrol and diesel engines in a wide range of applications including road vehicles and even tractors.

British companies had a considerable lead in the commercial development of gas turbine technology, including turboprop airliners and turbine powered ships. The development which caused the biggest sensation was the first gas turbine powered car demonstrated by the Rover company.

The Rover was quickly followed by experimental cars from other leading European manufacturers including Austin, Fiat and Renault. The 24-hour race at Le Mans had a special section for gas turbine cars which was dominated by the Rover company in an impressive display of speed and reliability.

American car and truck manufacturers also invested heavily in gas turbine research. The strongest commitment came from

***(Facing page)* Two experimental steam tractors built by International Harvester in the early 1920s.**

Chrysler with plans to build a batch of about 50 pre-production prototype cars, including some for use by members of the public, in an ambitious evaluation programme.

Fuel in a gas turbine engine is burned in a combustion chamber, heating air from an intake at the front. Burning fuel expands the air, accelerating it through the turbine blades and forcing the rotor of the turbine to spin.

One of the advantages of this type of power unit is the smoothness of the spinning rotor instead of a series of explosions and reciprocating pistons. Fast response times and excellent power-to-weight ratio are also attractions, and these helped to provide rapid acceleration when the turbine was used in cars. Another advantage of a gas turbine is its compact size, and this is useful in a car where the extra space can be used for passengers and luggage.

Gas turbines run on relatively low cost paraffin, but this advantage is outweighed by increased fuel consumption, which was one of the factors that helped to reduce the motor industry's enthusiasm for the gas turbine. Noise was another problem, and there was also concern about the lack of engine braking for vehicles which were expected to have a high performance on the road.

Several American companies were interested in the idea of using a gas turbine to power farm tractors, but International Harvester was the first to complete a prototype and provide a public demonstration.

An IH subsidiary company had designed a gas turbine for a helicopter, and this was the power unit used for the HT-340 tractor programme. The IH turbine and its reduction gearing weighed only 90 lb, and the 13 in. diameter and 21 in. overall length easily fitted into the engine compartment of a medium size tractor. The turbine was designed to deliver up to 80 hp, but it was restricted to 40 hp for the tractor because of the limitations of the transmission.

International Harvester was also developing a hydrostatic transmission for their tractors, and they were able to adapt this for the HT-340 tractor project. Gas turbines are most efficient when operating at constant speed, and this is an ideal application for a hydrostatic transmission. The HT-340 turbine powered a variable displacement pump delivering oil to radial motors built into the rear wheels, and this allowed the travel speed to be varied simply by adjusting the oil flow while maintaining a constant turbine speed.

The HT-340 attracted massive publicity, and this was probably one of the main reasons why International Harvester decided to develop the tractor.

A gas turbine offers few practical advantages in a farm tractor.

A wartime picture of a Renault with a methane gas kit.

A compact power unit giving rapid acceleration is not necessary in a tractor, and the smoothness of a turbine is less significant in a tractor than a car. Higher fuel consumption and noise are obvious disadvantages, and a gas turbine is more expensive to manufacture than a diesel engine of similar power.

Interest in gas turbines for cars waned, and International Harvester decided to abandon their turbine tractor project.

Engineers were interested in alternative fuels long before the first petrol powered cars and tractors appeared. Demand for agricultural steam engines was limited in some parts of the world by the lack of cheap coal, and the development of wood and straw burning engines helped British manufacturers to build up valuable export markets in areas where coal supplies were limited.

The rapid commercial development of the petrol engine also encouraged interest in fuel alternatives. Extra demand was pushing up the price, and there was concern about the long term availability of oil derived fuels.

Alcohol attracted particular interest as an alternative to petrol.

Spark ignition engines needed little or no modification to use alcohol, and the fuel could be produced almost anywhere from crops such as potatoes or cereals.

Potatoes were being grown on a substantial scale for alcohol production in both France and Germany by the end of the last century, and a statement issued by the Gardner Engine Company of Manchester in 1903 advocated a similar production programme in Britain. This followed the international trials of alcohol powered engines held in Paris in 1902 by the French Ministry of Agriculture, where one of the top awards had been won by a Gardner engine.

A motor plough powered by an engine designed to burn alcohol was available in Germany in 1900, according to a report in the British publication *Implement & Machinery Review*. It was made by the Oberursel Motor Factory in Frankfurt and was called a Plough Locomobile. The work rate was said to equal a steam plough, but the operating costs were lower because alcohol was much cheaper than coal.

Alcohol produced from potatoes cost the equivalent of 8p per gallon or 1.8p per litre in 1900, and the 20 hp Plough Locomobile used 11.4 litres an hour during a ploughing test.

Further information about the cost of alcohol for tractor engines came in 1905 when Dan Albone organised a fuel test with one of his Ivel tractors. He was interested in the possibility of alcohol from potatoes as an alternative to imported petrol, and the ploughing test was designed to show that tractors would work on alcohol and to measure work rate differences.

Petrol, paraffin and alcohol were used in the trials, and in each case the area of land ploughed with two gallons of fuel was measured. The results expressed in roods (0.25 acres) and poles (40 poles = 1 rood) were: petrol—3 roods; paraffin—2 roods 35 poles; alcohol—2 roods 25 poles.

When Dan Albone was concerned about fuel costs the price of petrol was equivalent to 6.4p per gallon. This seems cheap in present day terms, but it was expensive in 1905 when the wages of an unskilled agricultural worker were equivalent to about 1.4p per hour.

Another fuel alternative is producer gas, consisting of carbon monoxide, nitrogen, methane and hydrogen. It can be used as fuel for a spark ignition engine, and it can also replace up to about 70 per cent of the diesel fuel in a compression ignition engine.

The gas is produced when a stream of air is passed through hot carbon. The carbon can be in the form of coal, but producer gas installations on tractors are usually fuelled by charcoal or, in tropical countries, by-products such as coconut shells.

The gas production process takes place in a specially designed container which can be big enough to supply the gas to run a

Renault AFM-H crawler tractor working in 1943 with a gas conversion.

large stationary engine in a factory, or small enough to be carried on a vehicle such as a car, a truck or a tractor.

Producer or suction gas is unpopular because its energy value is less than petrol or diesel, causing a substantial reduction in the power output. Gas from some of the early kits often contained impurities such as tar and ash, and this caused reliability problems and could also reduce the working life of an engine.

The production kit is bulky enough to restrict the view from the driving seat, and the burner needs stoking with fresh charcoal every hour or so to maintain the gas supply.

In spite of these and other disadvantages, producer gas has often been used as an engine fuel. Charcoal is the usual source of gas for the units fitted to tractors, and the raw material can be tree trimmings and brushwood which are available almost everywhere and can be used when supplies of oil derived fuels are not obtainable.

The strategic significance of a home produced fuel supply was appreciated in France where the Academy of Agriculture and the army helped to encourage improvements in the design of

A modern gas conversion kit photographed on an International Harvester tractor at the Paris Show.

producer gas kits during the 1920s. When ordinary engine fuels were in short supply during World War Two, French companies were able to supply large numbers of producer gas kits for priority users, including farmers. The kits were also produced in large numbers in other countries, particularly Germany where they made a major contribution to providing tractor power during the war.

The kits became a familiar sight on buses, trucks, cars and tractors in Europe, and they were also used in substantial numbers in other areas of the world affected by the wartime fuel shortages. Thousands of tractors were working on producer gas in Australia, and there were also large numbers on farms in New Zealand.

There is still a demand for producer gas units for tractors. The kits are made in small numbers in France for export to parts of Africa where diesel fuel is difficult to obtain, but supplies of wood suitable for charcoal production are available. A kit suitable for a modern 80 hp diesel tractor costs about £3000. The starting up time for the gas production process is about ten minutes with modern equipment, and ash removal takes another ten minutes or so each day.

6

Ideas that Failed

THE tractor industry has produced plenty of success stories, but these have been outnumbered by a steady stream of failures. Many of the failures were the odd and impractical ideas that nobody wanted to buy. Some came from engineers with little or no practical knowledge of farming, and some were produced by farmers who knew little about engineering.

There were also some failures which simply arrived before the market was ready for them. These often became successes when they appeared again a few years later, and they include some of the important features in modern tractor design.

With such a high failure rate, developing new ideas for farm tractors is a hazardous occupation which only occasionally provides substantial rewards for inventors. Harry Ferguson's contribution to tractor history made him wealthy and famous, but he was one of the exceptions and Professor Scott's story is more typical.

Few people in the 1990s have heard of John Scott, but in the 1890s he was well known as the Professor of Agriculture at what is now the Royal Agricultural College at Cirencester. He had a

Professor Scott's 1903 tractor and powered cultivator.

particular interest in farm mechanisation, and when he left the college and returned to Scotland he began to develop his ideas for an 'agricultural motor' or tractor.

In 1897 he filed a patent for a petrol powered cultivator which was shown for the first time at the 1900 Royal Show, and that was the start of a remarkable series of tractors with features that were far ahead of their time.

His 1900 motor cultivator was equipped with a platform at the rear for carrying a small load, a feature that has appeared again more recently on transport tractors such as the MB-trac. A further development of his motor cultivator appeared in 1903 equipped with a rear mounted powered cultivator with vertical tines working on the same principle as the modern Lely Roterra cultivator. This machine also had a seed drill attached to the rotary tiller to allow seedbed preparation and drilling to be completed in a single operation—an idea that has recently become popular.

Developments for 1905 included a steam powered tractor using ideas that Bryan Harvester and International Harvester were rediscovering in America 20 years later (see page 94). The Scott steamer used paraffin burners instead of coal, with a lightweight tubular boiler enabling steam to be raised within three minutes of a cold start, according to the Professor.

The most surprising feature arrived on another new Scott tractor which was shown for the first time in 1904. This was a power take-off operated by a gear drive from the engine, and Professor Scott decided it should be at the front of the tractor where it could be used to power equipment such as a binder or a mower pushed by the tractor.

The 1904 model had three wheels, with a single rear wheel, and it was equipped with a device to raise the height of one of the front wheels to keep the tractor level while ploughing with the other wheel in the furrow bottom. It was also possible to vary the front wheel spacing for row-crop work, and the weight of the tractor was kept as low as possible to minimise soil compaction. The price was £200.

Surprisingly little information about Professor Scott's tractor business has survived, but the indications are that it was not a success and it was probably closed down in about 1906. He may have sold a small number of his tractors overseas, but the market in Britain was virtually nonexistent and his sales total was almost certainly small.

In spite of the lack of interest, Professor Scott remained convinced that tractors would eventually replace horse power.

'The time seems fast approaching when motor power will be universally used on the farm, and farm labourers will know more about motors than about horses,' he said at a farmers' meeting in 1906. His audience, in a part of Scotland where many farmers

Gougis tractor with power take-off in 1906.

bred working horses for sale, remained sceptical. Some objected to the smell of tractors, some complained about the dust they caused and others said tractors were useless on stony land.

After Professor Scott had failed to interest farmers in his power take-off, the idea reappeared in France on a Gougis tractor in 1906. It was used to power a binder, and this time it was at the rear of the tractor.

The Gougis p-t-o made little more impact than the Scott version had achieved, and the idea was not accepted until 1918 when International Harvester offered a p-t-o on their 8-16 Junior tractor. The IH power take-off was designed to operate at 540 rpm, which was later adopted as the standard p-t-o speed.

Tractors equipped with a set of reins or lines instead of the usual controls were made by several companies in America and Britain during the period when there were still large numbers of farmers switching from horses to tractors, but the idea does not appear to have been popular.

One reason for using a set of reins was to make it easier for inexperienced drivers to become accustomed to controlling a tractor. As most drivers had previously worked with horses, it may have been reassuring to be given a set of reins to hold.

Reins also made it easier for farmers to continue using existing horse-drawn equipment after they bought their first tractor. The driver could sit on the seat of the implement and use the reins as a long distance method of operating the tractor controls.

The peak year for launching tractors controlled by reins or

The first tractor to establish the idea of a power take-off was the International Harvester 8-16 Junior.

Petter's Iron Horse tractor was controlled by reins or plough lines.

lines seems to have been 1915. Among the newcomers were two American manufacturers, the Detroit Tractor Company with the Line Drive tractor and the Line Drive Tractor Co of Milwaukee. Both produced small, two-wheel tractors equipped with a set of lines to control the engine and the steering, and both had disappeared from the market within about two years.

Petter of Yeovil announced their rein controlled tractor at the 1915 Smithfield Show, and called it the Iron Horse. This was a three-wheeler powered by a 10 hp single cylinder paraffin engine. The rein control system was used to enable one man to sit on the implement and control both it and the tractor.

The Petter Iron Horse appears to have been as much of a failure as the American Line Drive models, but this did not deter others from using the same idea. The Samson Iron Works Co of Kansas City produced their line drive tractor in 1919 and, like the Petter company in England, called it the Iron Horse. The Samson four-wheel drive tractor was followed in about 1921 by the Adaptable motor plough from Chandler and Taylor of Indianapolis, Indiana also with a line drive system.

John Fowler of Leeds started making their Rein Control tractor in 1923. It was based on an Australian design, and incorporated

(*Above*)
Chandler and Taylor announced the Adaptable tractor with rein controls in 1921.

(*Right*)
The Fowler Rein Control tractor of 1923 was based on an Australian design.

a two-wheel power unit with an extendable drawbar for towing ploughs and other equipment. Driving the Rein Control tractor was 'much easier than controlling a team of horses', according to the Fowler publicity, as all the control functions could be operated with just one hand on the reins.

As the reins stopped and started the tractor, selected forward or reverse gear, applied the brakes and also steered the power unit to the left or the right, claiming that the controls were easy to use sounds like an exaggeration. The Royal Agricultural Society's machinery award judges confirmed the claim and were full of praise for the idea when they gave the Rein Control tractor a silver medal at the 1924 Royal Show.

'A slight pull at the left or right rein steers the tractor. A steady pull on both reins stops it, and a further pull puts the reverse gear into action. The reins are easily operated by one hand, leaving the other free to attend to the implement being drawn,' said the judges' official report.

The silver medal award was the Rein Control tractor's biggest success, as it was a failure in commercial terms.

A similar fate awaited some of the first tractors equipped with a power operated mechanism to lift and lower implements. Some manufacturers had developed this type of equipment long before the hydraulically operated Ferguson and John Deere versions turned the power lift into a success during the 1930s.

Some companies used a friction drive system to provide the lifting action but Bumsted and Chandler chose a rotating cam to operate the power lift on their Ideal tractor announced in 1912. Implements for the Ideal had to be attached to a special framework on the end of a chain linked to the cam, and the action of the cam raised the framework and the implement.

The mechanical lift system was just one of the features on a tractor of considerable technical interest. It was equipped with a differential lock, and the driving wheel spuds were self cleaning (see chapter 2).

An improved version of the Ideal was announced in 1917, with the engine power raised from 24 to 35 hp, larger diameter driving wheels and two braking systems—one operated by a foot pedal and acting on the differential shafts, and the other with drums attached to the rear wheels and operated by a hand lever. The Ideal could also be supplied with a p-t-o, suitable for a binder or grass cutting equipment, the makers suggested.

The special features made the Ideal complicated and expensive, and it disappeared from the market by the early 1920s.

Emerson and Brantingham were making tractors in America with a mechanically operated implement lift, while Bumsted and Chandler were doing the same in Britain, and the Tourand-Latil power lift tractors were being built in France. Emerson and

Ideal tractor with a power operated implement lift in a 1917 photograph.

(Right)
Tourand-Latil tractor built in France in about 1920 with a power operated implement lift.

(Below)
John Deere provided an hydraulic power lift mechanism on the Model A tractor in 1934.

Brantingham used friction drive from a groove in the flywheel, and they persevered with the idea longer than most of their rivals. Power lifts were attracting real interest in America by the early 1930s when John Deere was offering a mechanical lift, paving the way for the first hydraulic system in 1934.

Being too late with a new development can be even worse than bringing it on to the market too soon. This was the fate of the new cable ploughing tractors produced by three British manufacturers over a ten year period from 1913.

Cable ploughing with steam engines was developed in Britain during the nineteenth century. The aim was to keep the weight of the engines and winding gear on the headland in order to minimise soil compaction while the field was ploughed or cultivated. The equipment was expensive to buy and to operate, but work rates could be high and a cable set with a pair of ploughing engines could do as much work as a large number of horse drawn ploughs. Cable systems operated by contractors and large farms covered a big acreage each year in the arable areas of countries such as Britain, Germany and Hungary.

The development of smaller, lighter tractors should have made cable ploughing obsolete, but some companies believed the internal combustion engine would give the cable plough a new lease of life. Fowler, easily the biggest manufacturer of steam ploughing equipment, was the first to make a petrol/paraffin version.

When the first petrol/paraffin model appeared, it looked very much like one of the steam powered Fowler engines, a fact that was criticised by the Royal Agricultural Society machinery award judges for not breaking away from steam engine styling. The new tractor appeared in 1913, based on an earlier version built in 1911 without a cable drum. Both were equipped with a four cylinder engine developing 50 hp.

A new look cable ploughing tractor arrived from the Fowler factory in 1920. It was equipped with a 60 hp engine, and it was followed by a 100 hp version in 1922. A statement published in 1922 explained that the Fowler company believed steam was still the ideal power for cable ploughing, but the paraffin powered tractors, they said, would bring the benefits of cable systems to areas where unsuitable water supplies or lack of coal meant steam equipment could not be used.

The biggest competition for Fowler equipment as the cable ploughing market faded away was Walsh and Clark, also based in the Leeds area.

Walsh and Clark Victoria cable ploughing engine.

McLaren motor windlass built in 1920 for cable ploughing.

Walsh and Clark built their first paraffin powered engines for cable ploughing in 1913 and continued making them in small numbers until the early 1920s. Their Victoria engines, like the earlier Fowler models, had steam engine styling. Although this had brought criticism when a Fowler engine was entered for a machinery award, the Walsh and Clark steam engine look-alike won a Royal Agricultural Society silver medal in 1915 without, apparently, attracting criticism. Walsh and Clark used the space inside their boiler as a fuel tank, and claimed it held enough paraffin for four days' work.

An improved version of the Victoria was announced in 1918 with a twin cylinder paraffin engine mounted on top of the fuel tank. The maximum power output was 35 hp at 600 rpm, and the output from a set of two Victoria engines working with a four furrow balance plough was claimed to be up to ten acres a day. Cable ploughing with two engines requires a team of at least three operators, including one riding on the plough.

J & H McLaren—yet another Leeds based company—built their first cable ploughing tractors in 1920, presumably expecting the market to expand.

The McLaren tractor was called a motor windlass, and it was not designed to look like a steam engine. A simple chassis made of steel girders carried the four cylinder petrol/paraffin engine and a vertical winding drum. The engine developed 40 hp on petrol and 32 hp when running on paraffin.

Performance figures claimed by the manufacturer included fuel consumption at three gallons of paraffin an hour, and the work rate was claimed to be 1.43 acres an hour from a ploughing set consisting of a pair of motor windlasses, a 450-yard cable and a four-furrow balance plough.

One of the advantages claimed for the motor windlass was that

Fiat 602 tractor working in Turkey in 1920 with improvised weather protection for the driver.

of simplicity. Anybody 'with a working knowledge of a motor bicycle or a small petrol engine' could operate the equipment, so the company said.

The market for new cable ploughing equipment virtually disappeared during the 1920s. Most of the remaining customers had maintained their traditional preference for the steady pull of a steam engine, and the demand for paraffin powered engines remained small.

During the first 75 years of tractor development the only concession to comfort for most drivers was a bare metal seat mounted on a sprung support. On some tractors there was also a roof or a canopy to protect the driver from the midday sun and the worst of the rain, but this was generally reserved for big, expensive tractors.

The first serious attempt to provide farm tractor drivers with a really high standard of comfort came in 1938 when the Minneapolis-

Moline UDLX tractor was announced. The new tractor followed a market research survey in which more than 50 per cent of the farmers questioned said they would like a tractor with a cab.

'Just as the city man needs a comfortable closed car to pursue his activities, so the farmer who spends a big share of his time on a tractor needs and wants greater comfort on the job,' said the UDLX sales leaflet.

The company chose the word 'Comfortractor' to distinguish the UDLX from ordinary tractors, and confidently described it as 'THE WORLD'S GREATEST TRACTOR'.

The UDLX tractor was equipped with a five-speed gearbox, allowing up to 40 mph in fifth gear. Safety equipment provided in the standard specification included a self-energising Bendix brake system, safety glass in all windows, electric windscreen wipers, plus front, rear and brake lights.

The cab on the UDLX was a steel structure with insulation to reduce the noise level, and with two doors at the rear. It was big enough for two fully upholstered seats, and the list of equipment for the driver and passenger included an electric clock, a cigar lighter and a fitted ashtray. There was also a fitted radio and a heater—still regarded as up-market items in late 1930s cars and virtually unknown in the tractor market. A sun visor, rubber floor mats, an instrument panel light and a roof light were also provided.

Minneapolis-Moline used a four cylinder petrol engine in the UDLX. This was their KED series engine which was designed to burn leaded fuel and, in a less environmentally aware age, it was claimed as a major sales feature for the UDLX. The rated power output was 42.6 hp on the belt and 39 hp at the drawbar.

About 150 UDLX tractors were built during the two year production run, and many of these were bought as industrial tractors. Few farmers were willing to pay $2155 for the well equipped UDLX when they could buy two ordinary 40 hp tractors for the same money.

7

The Future Viewed from the Past

TRYING to guess future developments in tractor design can be difficult, but the engineers and stylists of the 1960s who showed their ideas for the future sometimes came up with the right answers.

Companies in the motor industry are often willing to invest large sums to produce futuristic one-off cars. These may be designed to check reaction to ideas which could be included in a future production model, but the most exotic cars of the future are built for the publicity they attract.

Tractors of the future have appeared less frequently, probably because the cost is more difficult to justify in an industry where

The futuristic Typhoon II tractor from Ford.

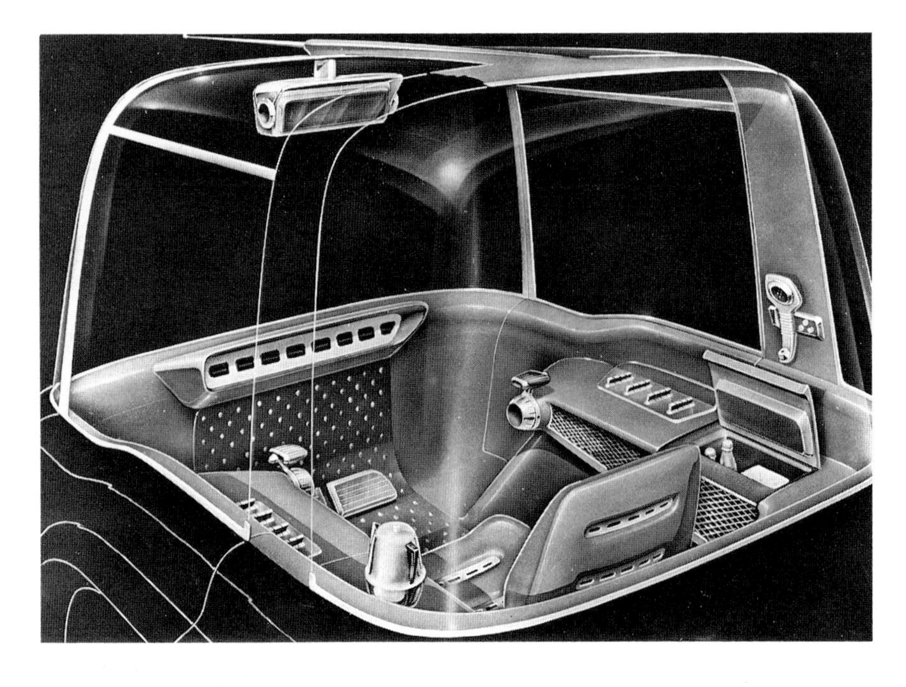

Detail of the Typhoon II cab interior.

profit margins are often slim and styling changes are less significant; the last batch arrived in the 1960s.

Ford cut the cost of their Typhoon II tractor by producing it as a half-size scale model. It was built in 1965 when tractor cabs were either primitive or nonexistent, but the Typhoon design team predicted future cab developments for up-market tractors with considerable accuracy, providing the driver with a radio, heater, air conditioning and a telephone link.

There was also a cool box for carrying drinks or the driver's lunch, and this is a feature which was beginning to appear on some models about twenty years after it was featured in Typhoon II.

The Ford designers provided their tractor with four-wheel drive through equal size wheels, and included four-wheel steering. They also provided four power take-offs—one at the rear, one at the front and one at each side.

With the engine placed at the rear of the tractor and the cab at the front, the driver in the Typhoon would have had good forward visibility but a less satisfactory view of equipment at the rear of the tractor. Ford overcame the lack of rear visibility by using a miniature television camera mounted behind the tractor and a screen in the cab showing a picture of trailed or rear-mounted equipment.

Kubota provided three external TV cameras on their Talent 25 tractor which was built in 1969 for display in the following year at the World Fair in Japan. The driver was able to switch from one camera to another to show pictures on a monitor screen of equipment behind the tractor and also views at each side.

Kubota's Talent 25 tractor with three external TV cameras.

Some of the Talent 25 design features appear to be impractical, and the tractor was probably produced to attract publicity rather than as a serious attempt to evaluate new ideas. The styling was obviously eye-catching, but it could have limitations on a farm. The curved body panels over the wheels would soon be plastered with mud and damaged by stones from the tyres, and there appears to be little space around the wheels for a build-up of mud or to allow a sharp steering angle.

One of the least practical features was the use of two rear doors to enter or leave the cab. Even the most nimble of drivers must have found this difficult with an implement mounted on the rear linkage, and the rear view was obstructed by the edges of the closed doors.

A more forward-looking design feature was the swivelling seat. This could be turned through 180 degrees to allow the driver a better view when using equipment such as a rear-mounted forklift attachment. It may also have been provided to make it easier for the driver to enter or leave the seat from the rear doors. Kubota later became the first company to offer a reversible driving seat as an option on a compact tractor.

The external TV cameras providing a view to the rear of the Kubota tractor and the Ford Typhoon II were 20 years or more ahead of their time. In 1990 SAME offered similar equipment as an extra cost option on some of their production tractors, showing pictures from the rear view camera on a mini-sized screen built into the instrument panel behind the steering wheel. Closed circuit television has also been used on equipment such as potato harvesters, allowing the operator to check the operation of areas on the machine which are normally out of sight.

International Harvester had two principle objectives when they built the HT-340 tractor in 1961. It provided a test vehicle for evaluating a gas turbine power unit and hydrostatic transmission (see chapter 5) and it was also designed to attract maximum publicity. It is one of the most innovative and interesting experimental tractors the industry has produced, and it has been on display in the agricultural section of the Smithsonian Institution in Washington, DC.

Sleek streamlining was the fashion in the motor industry in the early 1960s, and the International Harvester styling team followed the trend with the HT-340—even though a wind-cheating body shape is not particularly useful on a slow moving farm tractor.

The shape helped to encourage publicity, but a more interesting idea was the use of fibreglass to make the HT-340 bodywork. Fibreglass has a high degree of corrosion resistance, and the colour can be moulded in to give a longer-lasting finish. Body panels made of various synthetic materials have been used by some tractor manufacturers since the early 1980s, but sheet steel and paint are still the standard materials.

In spite of including so many forward-looking features, the design team apparently overlooked two of the most important developments for the future. The HT-340 was not provided with a cab, and it was not equipped with four-wheel drive.

The HT-340 tractor featured gas turbine power and glass fibre body panels.

United States Steel published details of their ideas for a new farm mechanisation system in 1970. It was based on the Vantage tractor and a range of more than 25 specially designed machines and implements. The aim, according to a booklet describing the project, was to stimulate interest in tractor design and to encourage the development of more advanced equipment for agriculture.

'Though bold and unusual by today's standards, it is a realistic approach to the agricultural industry's needs of the seventies,' the booklet explained.

The tractor unit was based on a V-8 engine providing up to 225 hp at the p-t-o. The design team had decided that this would be the most popular power output for large agricultural tractors in America during the mid-1970s, rising towards 300 hp by about 1980. The designers also specified a hydrostatic transmission with four speed ranges, using individual wheel motors to provide four-wheel drive.

Vantage was also designed with four-wheel steering. The driver could select one of four modes, steering with the front wheels, the rear wheels, all four wheels or crab steering. The tractor was equipped with front as well as rear implement attachment points complete with a quick hitch system, anticipating a 1980s trend in European tractor development. The power take-off points were at the rear of the tractor and built into the right-hand side. It was suggested that a long drive shaft with two 90 degree drive joints could be used to take power from the side p-t-o to front mounted implements, a complicated alternative to including a front p-t-o in the original design.

Driver comfort and safety were not high on the power farming priority lists in the early 1960s, but this is one of the areas that the US Steel team considered carefully and imaginatively when they were designing the Vantage tractor.

One example of this was the decision to include a four-wheel braking system to give increased stopping power on the road. Twenty years after details of the Vantage project were published only a small proportion of new tractors were equipped with four-wheel braking, although conventional two-wheel braking systems were often described as inadequate.

The Vantage cab was designed for strength and safety, and was part of the main structure of the tractor. There was also detailed attention given to the position of the grab handles and the design of the steps giving access to the cab, with non-slip surfaces provided.

Another development anticipated by Vantage was the use of external controls for the implement hitch. The driver could stand beside the rear or the front hitch and use push-buttons to operate the hydraulics and make hitching-up easier and faster.

Although Vantage was never built and existed only on paper, it was a remarkably detailed and practical attempt to suggest how big tractors might develop.

One of the developments the designers of Vantage and other tractors of the future were unable to anticipate was the impact that new technologies, particularly electronics, would have on tractor design. Tractor electronics were almost unknown in the early 1980s, but within ten years most of the leading manufacturers were offering electronic information and control systems on at least some of their tractor models.

Electronic systems can collect and process information quickly and precisely. The information can be displayed in the cab to help the driver decide which adjustments to make, but some electronic systems also have a control function which takes over some of the decision making from the driver.

Information systems include the fuel economy displays which first appeared on some Renault and Steyr tractors. These are based on information collected from the tractor engine and transmission, and the display panel inside the cab informs the driver when a different gear ratio or changing the engine speed would improve fuel economy. The driver can decide to make the adjustment and save fuel, or the information may be ignored if the driver decides that maximum output is more important than fuel economy.

Another stage in the development of information systems was announced by Massey-Ferguson in 1990 when they showed a prototype version of a data transfer system. Performance information collected and stored on a mini computer inside the tractor cab can be transferred at the end of each day to the main computer in the farm office for permanent storage.

Electronic controls began taking over some tractor driving functions in the mid-1980s. An electronic control system using information collected by sensors in the rear linkage can adjust the working depth of a plough in response to changing soil conditions more precisely than even a skilled tractor driver. Another example of electronic control is the engine management system demonstrated in Britain in 1987 by Lucas-CAV, which showed that electronics could increase work rates for jobs such as ploughing, and also achieve more precise speed control to increase the accuracy of spraying and fertiliser spreading.

Not surprisingly, those who were predicting future trends during the 1960s failed to forecast the use of satellite navigation systems for farm tractors, a development which began on an experimental basis in 1989, a century after the first tractor was built.

Satellite navigation systems use signals to pinpoint the location of ships and aircraft, and the same system can be used to show

United States Steel designed the Vantage tractor for the 1970s.

the position of a farm tractor or combine harvester.

The equipment can show the whereabouts of a tractor or a machine with an accuracy of just two or three metres, and the idea is to use satellite signals to plot the position on a map of the field. Navigation equipment installed on a combine harvester would allow a mini computer in the cab to record yield information and the position of the combine on a map of the field. This information could help to identify problems such as compaction, areas of poor drainage, disease or a mineral deficiency causing reduced yields.

The yield map prepared by the combine harvester could also be fed into a navigation system on a tractor. It would be used to adjust fertiliser or spray applications automatically for different areas in each field, reflecting the variations in yield identified by the combine harvester. Much of the pioneering work on satellite navigation in tractors and combines came from Denmark, a country which has led the way with legislation to control the use of agricultural chemicals.

As the farming industry faces increasing environmental pressures, fertilisers and spray chemicals will have to be used more precisely and more sparingly. This, as well as economic pressures, could take tractors into the space age with a satellite link to help achieve the extra precision demanded in a greener world.

Index

Page numbers in *italics* refer to text photographs. Numbers in **bold** refer to the colour plates, which are between pages 64 and 65.